HOW WORK & CONFIRM DX – THE EASY WAY

By: Craig E. "Buck," K4IA

ABOUT THE AUTHOR: "Buck," as he is known on the air, was first licensed in the mid-sixties as a young teenager. Today, he holds an Amateur Extra Class Radio License.

Buck stands on the ARRL Honor Roll and has earned the 8 Band DXCC award. He is also an active instructor and a Volunteer Examiner. The Rappahannock Valley Amateur Radio club named him Elmer (Trainer) of the Year three times.

Email: K4ia@EasyWayHamBooks.com

DX – The Easy Way and other Easy Way Books by Craig Buck are available at Ham Radio Outlet stores and Amazon.
How to Get on HF – The Easy Way
Prepper Communications – The Easy Way
Pass Your Amateur Radio Technician Class Test – The Easy Way
Pass Your Amateur Radio General Class Test – The Easy Way
Pass Your Amateur Radio Extra Class Test – the Easy Way
Pass Your GROL Test – The Easy Way

Copyright ©2016, Craig E. Buck   All Rights Reserved. No part of this material may be reproduced, transmitted or stored in any manner, in any form or by any means without the express written permission of the author.

Latest revisions February 2021   V1.10

ISBN 978-1523286645

DX - THE EASY WAY
TABLE OF CONTENTS

TABLE OF CONTENTS	1
INTRODUCTION	4
HOW I GOT STARTED	5
THE HARDEST ONE	12
WHAT'S A COUNTRY?	15
CASUAL DX	19
RARE DX	22
VERY RARE DX	25
DX CODE OF CONDUCT	29
DX ETIQUETTE, DECORUM AND PROPRIETY	37
OPERATING TIPS AND STRATEGIES	41
TRAIN YOUR EARS	41
ZERO BEAT	42
LISTENING	42
USE PASSBAND TUNING	44
REDUCE YOUR RF GAIN	44
OPERATING SPLIT	45
TIMING	47
PHONETICS	48
CW STRATEGIES	50
GET A NEW CALLSIGN	51
ANATOMY OF A QSO	52
PROPAGATION	57
PREDICTION PROGRAMS	57
BEACONS	59
REVERSE BEACON NETWORK	60
GENERAL CONSIDERATIONS	60
CALL CQ	61
MODES	64
SINGLE SIDEBAND (SSB)	64
CW (MORSE CODE)	66
DIGITAL MODES (PSK, FT8, RTTY)	69

WHO'S ON?	72
MAGAZINES	72
ONLINE AND BULLETINS	72
SPOTTING NETWORKS	72
DX FREQUENCIES	75
TRANSCEIVER SELECTION	77
TRANSMITTER	77
PANADAPTERS	80
AMPLIFIERS	82
ANTENNAS	85
VERTICALS	85
HORIZONTAL DIPOLES	86
ROTATABLE ANTENNAS	86
SHACK DESIGN	88
COMPUTERS	92
WHERE CAN YOU OPERATE?	94
CONFIRMING YOUR CONTACTS	97
QSL CARDS	97
QSL DIRECT	100
OQRS	104
QSL VIA A MANAGER	105
QSL VIA BURO	107
QSL SERVICES	109
LOGBOOK OF THE WORLD (LoTW)	110
COMPUTER LOGGING PROGRAMS	113
CONTESTS	121
AWARDS	124
FISH STORIES – THE ONES THAT GOT AWAY	126
BE READY	126
BE QUICK	127
BE CURRENT	129
LEARNING CW	131
SUMMARY & A FINAL THOUGHT	135
DX ENTITY LIST	136

INTRODUCTION

I found myself getting more and more excited while gathering my outline for this book. I was even more passionate about DX than I imagined, and I hope to convey that ardor to you.

This book is for the HF DX chaser. It is not advice to DXpeditioners. I endorse no particular manufacturer or product. On occasion, a product might be mentioned to describe its functionality. Read the reviews on eHam.net, *QST* magazine, and other sources online and in print for comparisons.

I offer my suggestions on these pages, and there may be differences of opinion or some qualifier I missed. This leads me to point out a truism I discovered a long time ago:

> "If you ask five hams a question, you will get seven different opinions and maybe a fistfight."

Please send your comments, corrections, and chastisements to me at K4ia@EasyWayHamBooks.com

HOW I GOT STARTED

Begin at the beginning. I grew up in the days before personal computers, the internet, video games, and 500 channels of color TV. We made our entertainment. I remember building an airplane cockpit using shoe boxes with toothpicks for control levers. We shot marbles and created elaborate obstacle courses that would make a golf course designer proud. If ants got in the way, we dumped lighter fluid down their hole and lit it. No mercy. That was third grade.

These were the days when you rode a bike without a helmet; cars didn't have seatbelts, and you stayed out until the street lights came on. We played in the woods, caught snakes, and lit cherry bombs. Model boats and planes filled my shelves. In fifth-grade science class, we built a crystal radio set. It only tuned one station. At night, I would tuck the earpiece under my pillow and fall asleep listening to basketball games. Basketball on the radio took a lot of imagination and held no interest for me. I was just amazed to hear sound from a radio with no batteries. The radio's main component was a rock. That station was about 30 miles away.

Tuning the family's Crosley 5-tube AM radio, I discovered "skip" –radio waves bouncing off the ionosphere. I was excited to hear the big 50,000-watt AM clear-channel stations booming across the country. That was nothing compared to the excitement to come.

That's how I became a fledgling DXer. I guess I have to credit my parents for kicking it up a notch. When I got older, dad told me they tried to stimulate creativity when selecting our Christmas presents. I remember model airplanes and ships. One year it was a chemistry set, and the next, an Erector set. I think I

HOW I GOT STARTED

was in 7th grade when a Heathkit shortwave receiver appeared under the tree. I was already a builder, so I looked forward to the assembling and soldering, but nothing prepared me for what I would hear come through that speaker.

Miraculously, the radio worked. There was something eerie about hearing Radio Havana Cuba or Radio Moscow during the Cold War. I kept expecting the FBI to break down the door. I got my first lesson in political "spin" hearing Cold War propaganda even before the term spin was invented. I remember thinking it was funny that the tune used to identify Radio Havana was the same one the Gillette Razor Company used in their commercials. I guess since Castro had a beard, he didn't know about Gillette razors.

Recently, I got this card from Russia. It brought back memories of those days. His note thanked me and asked for my QSL card because he needed Virginia for his WAS (Worked All States) award. How times have changed.

I didn't become an airline pilot, shipbuilder, chemist, or architect, but that Heathkit radio was my introduction to real DX and, eventually, Amateur Radio.

HOW I GOT STARTED

The Cold War was in full swing, and there was a battle for the hearts and minds of the world carried out on shortwave radio. Voice of America, Radio Free Europe, and the BBC competed against Radio Moscow and Radio Havana Cuba. Virtually every country had a shortwave service, and most had some English-language programming. I didn't understand or care what they were talking about, but it was exciting to hear a signal from another part of the world. There was Deutsche Welle, the voice of Germany, HCJB from Quito, Ecuador, and many more I've long since forgotten. My young imagination led me on travels around the globe. The seed was planted listening to distant stations, fading in and out of the static, sometimes with distinctive polar echoes, but always conveying a sense of magic.

Tuning across the bands, I heard people talking to each other. That was how I discovered Amateur Radio. Immediately, I knew that if I became a Ham, I could graduate from a passive listener to a participant.

We were experimenters and builders – kids with curiosity. Today, we would be called "makers." What about ham radio? Other than DX, what was the lure? Simple. Ham Radio was the first time all the stuff they were trying to teach me in school had a use. How many times have you heard adolescents complain, "I don't need to know algebra." "I'll never use chemistry." "Why do I have to learn about the ionosphere?" "Who cares where Pitcairn Island is?" And, on and on.

If you want to know how long to make an antenna, use a little algebra. If you want to know the best time to talk to Australia, understand the ionosphere. Battery chemistry will help decide the best power source for your equipment. Knowledge of geography will tell which way to turn a directional antenna. If you know languages, you can say "hi" to a foreigner in

HOW I GOT STARTED

his. Astronomy helps one appreciate meteor scatter and sunspots or how to bounce a signal off the moon or satellite. Use physics to design an antenna. Knowing history, you can appreciate Pitcairn Island. I could teach most of an entire high school curriculum based on Amateur Radio alone.

And what does Pitcairn Island have to do with it? Plenty. During the summer, Mom would take me to the library. I enjoyed the tall-ship swashbuckling tales with pirates and mutinies. My favorite was The Mutiny on the Bounty – a true story from 1789 with real people, a real mutiny, and real villains and heroes. It is up to you to decide who the villains were and who were the heroes. The mutineers, led by Fletcher Christian, fled to Pitcairn Island in a remote part of the Pacific Ocean.

After I got my Ham license, I was tuning the bands late one night and heard VR6TC, Tom Christian, a descendant of Fletcher living on Pitcairn Island. VR6TC was Tom's callsign at the time. I "worked" him, and I was shaking.

Tom was famous because of the Bounty story and the rarity of his location. Pitcairn is remote and isolated. Tom and his wife had their licenses because there were no other means to communicate with the outside world in those days.

I wrote a "What I did Last Summer" essay about contacting Tom Christian for English class. My teacher didn't believe me. She said the composition wasn't supposed to be fiction. Months later, I received a QSL card from Tom. I enjoyed rubbing her nose in it. Maybe that's why she gave me a C in English that quarter. Gloating is never wise.

After college, I went QRT, and somehow, over the years, Tom's QSL card, VR6TC, was lost, but you can

HOW I GOT STARTED

see it on the wall to the top left edge of the map in my picture below.

I never heard Tom on the air again, and he became a Silent Key a few years ago. I have worked and confirmed Pitcairn many times since, but nothing can replace my Tom Christian card. My contact with Tom is one I will never forget.

K4IA as WA4TUF in 1967

I am not sure why I took that picture with a microphone. I hardly used one because my squeaky voice was a dead giveaway that I was "just a kid." On CW, no one knew.

HOW I GOT STARTED

Ducie Island, below, is an uninhabited atoll in the Pitcairn Islands group. Edward Edwards, the captain of *HMS Pandora*, sent to capture the mutineers of *HMS Bounty*, rediscovered Ducie in 1791. It became a separate DXCC entity in 2001. The first DXpedition was turned back by foul weather, but another attempt was successful in 2002, and so was I.

HOW I GOT STARTED

In the Holy Monastery of Docheiariou at Mt. Athos, Monk Apollo had his license for the same reason as Tom Christian – to maintain contact with the outside world. The monastery was founded in the 10th century and is cloistered, with no telephone service. When one of the monks fell ill, they could not call for aid.

Talking to Monk Apollo is another incredible Amateur Radio thrill. His monastery duties meant he was not on the air very often, but he was a cheerful QSLer. Sadly, Monk Apollo is now a Silent Key, but another monk has taken his place on the radio.

Here is Monk Apollo's card from Mount Athos. Mount Athos is a separate DXCC entity[1] from Greece, and this card is another of the rarer ones in my collection.

---

[1] More on "DXCC Entities" later.

THE HARDEST ONE

People who aren't Hams will ask, "How far can you talk on that thing?" They don't understand that distance is not the problem. Under the right conditions, you can work around the world – even the long way around. You might hear the echo of your signal or a station coming in by the short path and long path simultaneously.

Hams ask, "What was the toughest DX you ever worked?" For me, without a doubt, the answer is Bouvet Island.

Bouvet is an uninhabited volcanic island said to be the remotest spot on the planet. Located in the Southern Ocean, the nearest land is Antarctica, almost 1,100 miles away. The next closest is South Africa, 1,600 miles to the north-north-east. The island is only 20 square miles, and a glacier covers ninety-three percent.

Bouvet was discovered in 1739, rediscovered in 1808, and finally claimed by England in 1825. Over 100 years later, Norway landed an expedition and claimed Bouvet Øya (Bouvet Island) for Norway. After a short dispute, Britain gave up its claim, and the Island has been a Norwegian dependency since 1930. No one wanted it. Bouvet was declared a nature preserve in 1971, making it even harder to gain access.

Bouvet Island is number 2 on the DXCC[2] most wanted list and number 2 on the ClubLog most wanted list – only behind North Korea (which flat-out does not allow Amateur Radio at this time).

---

[2] DXCC is the DX Century Club – a program of awards administered by the American Radio Relay League.

THE HARDEST ONE

Norway operates a remote weather station from Bouvet, and since that requires periodic maintenance, therein lies hope for Amateur Radio activity. A DXpedition had been to Bouvet in 1989, but I was inactive at the time. It was another 11 years before Bouvet was on-the-air again. Don't ever let one of these rare opportunities get away! You might have to wait a long time for another chance.

In December of 2000, American Ham and space-walking astronaut Chuck Brady, K4BQW, activated Bouvet. He was there as a medical officer for a scientific team and played radio in his spare time.

Inclement weather and equipment malfunctions plagued the operation. Chuck couldn't run down to the local auto supply store for parts to repair the ailing generator. He re-hung the antennas daily, as they fell victim to 120-knot winds. The amplifier did not cooperate, and he was often reduced to 100 watts. To his credit, Chuck kept plugging away for just short of three months, making 40,000 contacts.

The path to North America was "only" 8,000 miles, and it was going to be a tough fight up against the hordes of other stations calling 3Y0C. I knew I had a shot, but could I break through the wall of Europeans?

I had to listen a lot if I was to be ready when the signal came. Unfortunately, I had a tough time hearing Chuck, and you can't work 'em if you can't hear 'em. I listened day and night but was beginning to despair. When I could hear him, the din of the pileups was overwhelming.

All rules went out the window with stations calling out of turn, ignoring the operator's instructions, and calling on frequency when Chuck was operating split. We'll talk more about all that in other chapters, but for now, know it was a challenging situation.

THE HARDEST ONE

The only band I heard Chuck on was 20-meter SSB, not my favorite mode and not my favorite mode for DX. I kept trying. Finally, on February 10, 2001, at 5:30 in the afternoon local time, I heard Chuck come back with, "K4IA 59." I did it! Here's the proof. How I did it is the subject of this book.

DXpeditions[3] to exotic tropical islands sound like fun. Operating from Bouvet looks like a nightmare. Chuck's twenty-two days in space must have seemed a picnic compared to three months of isolation on Bouvet.

Because of its remoteness, inaccessibility, rarity, and the fact this was a single operator fighting the elements and Murphy's Law, Bouvet rates #1 on my "toughest DX" list. It has been twenty years now. I wonder, how long before another activation? An attempt in 2018 was turned back by bad weather and a faulty boat engine. Keep listening!

---

[3] A DXpedition is an expedition to a DX entity organized by volunteers, sometimes at great expense and even danger.

WHAT'S A COUNTRY?

So you ask, "How does Bouvet, a remote uninhabited speck in the middle of nowhere, qualify as a country?" Good question. More accurately, we should refer to them as "entities," not countries.

The American Radio Relay League (ARRL) administers the DX Century Club (DXCC) program and has identified 340 entities as of this writing. You earn the DXCC award by confirming contacts with 100 entities. There are endorsements for more entities, for single bands, and for confirming 100 on five bands and more. The big prize is Honor Roll, which, currently, requires 331 confirmations. The ultimate goal of #1 Honor Roll demands confirmation from all current entities.

An entity is whatever ARRL says it is. Okay, maybe not as arbitrary as that sounds, but the rules are complicated. ARRL established the first DXCC list in 1935 with the statement, "The basic rule is simple and direct: Each discrete geographical or political entity is considered to be a country."

"Political entities" are those areas identified by government or political division. It must meet one of the following criteria:
1. Be a UN member state, or
2. Be issued a callsign block by the International Telecommunications Union (ITU), or
3. Contain a permanent population, administered by a local government at least 800km from its parent.

"Geographical separation entities" are those areas where a single political entity is separated into two or more parts. The area containing the capital is considered the parent entity.

WHAT'S A COUNTRY?

On land, if the area is separated from the parent by at least 100km of land belonging to another entity, it can count as a separate entity. Thus, Alaska is a separate entity from the US because Canada separates it from the mainland USA by more than 100km. Guantanamo Bay is a piece of the USA in Cuba, and it counts as a separate entity. Cuba owns Jose Marti Park in Tampa, Florida. The .14 acre park is officially Cuba, and it is more than 100km from the Island of Cuba. Why not Jose Marti? I can't answer that one.

Island areas count as a separate entity if the island is separated from the parent by 350km or more and also more than 800km from any other island attached to the parent. Thus, Hawaii is a separate entity from the US.

If intervening land or islands of another DXCC entity separate an island from its parent, it can count as a separate entity. Market Reef (OJ0) is split between Finland and Sweden. The Swedish side separates the Finnish side from Finland, so the Finnish side counts as a separate entity.

DX – The Easy Way

WHAT'S A COUNTRY?

It pays to watch world events. Upheavals bring changes to the political and island entities as well. When the former Soviet Union collapsed, Eastern Europe re-Balkanized and several new political entities arose from the former Yugoslavia and Czechoslovakia. When the Netherlands Antilles separated from Holland, individual islands became new entities.

When some group declares independence, it does not always create an entity. Kosovo waited ten years after it declared independence before finally being recognized in 2018. The critical point to remember is that an entity is not an entity because you think it qualifies. Only ARRL can designate an entity. It is nice to know the rules, but there is not a thing you can do about them.

Another class of entities is called "Special Areas." They count as a separate parent entity with special recognition. An example is the International Telecommunications Union headquarters in Geneva, Switzerland (4U1ITU), and the United Nations Building in New York (4U1UN).

WHAT'S A COUNTRY?

The Ham activity has to be legitimate and authorized by the government of the entity. You can't sneak into North Korea and start operating. ARRL does not recognize renegade operations for DXCC credit. The North Koreans might also take a dim view, and they have no sense of humor.

No country owns Antarctica, but it is host to scientific stations from different countries. As a result, Antarctic stations can have many different callsigns derived from the country of origin. For DXCC purposes, all stations in Antarctica are considered to be from one entity, namely Antarctica. Antarctica's "main" prefix is CE9, but this card is from the Russian research station at Vostok, so the callsign begins with R.

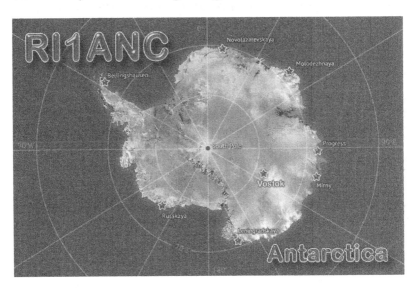

CASUAL DX

The first and easiest DX falls into the category I call "Casual DX." There are many easy-to-contact entities in Europe, Central and South America, the Caribbean, and Japan, with thousands of operators waiting for you. Looking at the DXCC "Most Wanted" list,[4] I would characterize 125 to 150 entities as Casual DX. You can easily earn your DXCC off this list using SSB and without ever slugging through a pileup. If you are willing to learn CW and digital modes, it will be even easier.

Don't assume every DX station is run by some hard-nosed contester who will chew you out if you make a mistake. Many of them are as green as you and may be operating with minimal stations. DX stations might operate from tiny yards or apartments with compromise antennas, low power, and antique or homebrew equipment. Your signal may be very loud compared to the signal you hear from them, so don't assume because a station is weak, you won't be able to work him.

DX stations need you for their Worked All States (WAS) award. Remember that Russian card a few pages back? There are quite a few county-hunters as well. I remember working an Englishman while driving through the Virginia countryside. He was ecstatic to get several new counties in his log.

Other DX operators might want to practice their English, and, despite what mean-scream media says, the good old USA is looked upon with favor by most of the world. The point is they are as anxious to make contact as you are. Get over your "mic fright" or "key clutch" and make the call.

---

[4] Several organizations survey and maintain a list of "most wanted" entities. They use different criteria but the results are reasonably consistent.

CASUAL DX

Most DX QSOs (contacts) are short, so don't worry about what to say. In the Casual DX QSO, exchange an honest signal report, QTH (location), and name. I call that the QSO trinity.

Proper etiquette is that the station who had the frequency controls the conversation. Listen to what he is doing. If he gives just a signal report, respond with that. If he wants more, he may provide the QSO trinity or offer information about his station, the weather, how long he's been a ham, etc. Then, you can reply with yours. If it is just a signal report, don't respond with more. His English could be very limited, or perhaps, he is just trying to make contacts. If he doesn't offer more, ask for a QSL card, thank him for the contact and get out of the way.

You can raise any category of DX by calling "CQ DX," and I'll have more to say about that later. Calling CQ DX works well with Casual DX because there are so many Casual DX stations. Give it a try.

On 40 meters, Europeans can transmit SSB in a lower portion of the band where Americans are not allowed. They will operate split[5] by calling CQ DX in their section and giving a frequency to listen in the US section. That's some fancy operating, but it can be very productive. Listen below 7.125 MHz, and you might be surprised.

To answer a Casual DX station calling CQ, QRZ or ending a conversation (be sure he is really done), you would give a short call to him like this: "His Call this is (or "de" on CW) Kilo 4 India Alpha, Kilo 4 India Alpha." Use the phonetics on your call because letters are hard to decipher amid noise and fading. See the section on

---

[5] Split is transmitting and receiving on different frequencies. More on that later.

CASUAL DX

Phonetics in the Operating Tips and Strategies chapter.

You don't need to use phonetics on his call. I am sure he knows his call. You don't need to be repetitive saying or keying his call several times and your call more than twice. He knows you are calling him because you are zero-beat[6] on his frequency (hopefully). Who else would you be calling? If he doesn't copy your call, he will say "QRZ?" or "Who is the K4?" and you try again.

Speak clearly and enunciate. Keep it simple. Avoid idioms, slang, complicated words, and long or tortured sentence structures. If you tell a non-English-speaking person the band is "dead as a doornail," he probably won't know what you mean. He won't know what a "bodacious" signal is either.

There are many Casual DX stations using digital modes such as CW, PSK, FT8, and RTTY. CW, FT8, and PSK have a considerable advantage over SSB in that you can copy very weak signals. You should add CW and digital to your operating modes, and I'll discuss that in the chapter titled Modes.

Chasing Casual DX can be very productive Have fun contacting new countries without the tension and frustration of deep hunting and pileups. However, as you get better, you will appreciate the excitement of hunting and working the rarer ones. No one wants to fish only for bluegills – especially once they learn of bigger fish in the ocean.

---

[6] Tuned to be on the exact same frequency he is using to transmit.

RARE DX

I would characterize the next 150 to 175 entities on the most wanted lists as "Rare DX." These are the entities not often heard. Maybe only a few hams live there, or the entity is only visited sporadically, or propagation to that part of the world is difficult for you. Usually, it is all three. You can reasonably expect to work them, but it is not going to be "casual."

The Rare DX station may be a laid-back operator who happens to be in a rare place. He may be no more experienced than you and his equipment no better than yours. He may hate pileups. He thinks sending QSL cards is an expensive chore. He has a license to communicate with the outside world, but becoming your DX puppy is not his plan. These laid-back operators are likely to respond to your CQ. They can enjoy a contact without fighting a pileup and then be gone. Many times, I have been surprised to get a call I never expected.

More often, you won't be that lucky. Rare DX takes some searching, and you are going to have to learn how to work "split." When a Rare DX station appears, it doesn't take long for the world to find out, and suddenly there can be hundreds of stations calling.

Just the other day, I heard a station in the Philippines (prefix DU) calling CQ on 40 meters around sunset. He was loud, so at first, I thought he was a "slim," [7] or maybe I had copied the call wrong. Surely the Philippines would not be coming in that loudly. Rule #1 – work them first and verify later. Listening confirmed it was indeed "DU." About that time, his call appeared on the spotting network[8], and bedlam broke loose.

---

[7] "Slim" as in slim chance, probably not real.
[8] Spotting networks allow stations to post what they are hearing to the

RARE DX

Everyone was calling on his frequency because he was operating simplex – transmitting and receiving on the same frequency. The problem with simplex is that the calling stations cover up the DX. You can't hear his response. "Did he call me?"

Fortunately, this operator knew what he was doing. He went silent and waited for a break in the action. Then he called "CQ de His Call Up Up." The "Up" meant he was listening above his transmitting frequency. Listening Up spreads out the callers and leaves the transmit frequency clear for the DX. "Up" means anywhere from 1-5 kHz above the transmit frequency on CW and 3-10 kHz on SSB. In big pileups, the spread can be greater.

It took a minute or two, but eventually, people stopped calling on his frequency, and I could snag the contact. Learn to work split to crack into the Rare DX level and beyond. I'll explain more about working split in another chapter, so leave the details for then.

Another vital skill is operating CW. Many DX stations use CW exclusively. CW hasn't been a requirement for US licensing since 2007, but it is still very much alive. Other chapters will discuss the advantages of CW over SSB and give hints on how to learn. I submit that only a rudimentary knowledge of CW is sufficient to work the Rare DX category. Let me explain.

Rare and Very Rare DX QSOs are usually no more than a signal report. Once again, the station that had the frequency dictates the protocol. If the DX is tearing through the crowd like an octopus on crack, the QSO trinity of Report, QTH, and Name shortens to only a signal report, and it is always 59 on SSB and 5NN on CW. 5NN is short for 599.

Internet so you can see what stations are on the band. More later.

RARE DX

You call the DX by giving your full callsign once. Notice, I said <u>full</u> callsign. Don't use his callsign at all. Listen for 5-7 seconds, and if he hasn't come back to anyone, throw out your full callsign again. If he calls you, say 59 (5NN on CW) and maybe "73" or "Thanks" ("TU" on CW) – not one word more unless the DX initiates it.

So, how much CW do you need to know? I already said there wouldn't be any conversation. Even if you don't "know" CW, learn to recognize the cadence of your callsign.

Don't rely totally on a DX spotting network to identify the DX, as the network may show a busted call. Always verify what you are hearing. Do this by recognizing enough CW to decipher the DX callsign. He will be giving it often (hopefully). Decode a letter at a time until sure. If in doubt, remember Rule #1: work first and verify later. Work them and verify the call by listening longer.

Learn to send your call and 5NN. If that is too much, program two buttons on a memory keyer[9]: one with your callsign and one with "5NN TU." Mash one button to send your call and the other to give the report when he calls you.

That's The Easy Way – and you don't need to be CW conversant. Try it. It works.

---

[9] Some transceivers and logging programs have a built-in memory keyer or you can get a separate box. Hit a button and it sends the message.

VERY RARE DX

The last 50 or so entities on the Most Wanted lists are tough. The last 25 are extra tough. Like Bouvet, these are places where amateur radio activity is sporadic or forbidden, such as North Korea. Your main concern will be waiting for regime change or someone to activate the entity, which can take decades. I call these "Very Rare DX."

Navassa Island was silent for 20 years until the latest DXpedition[10] went there in 2015. Navassa lies between Haiti and Cuba, so it is only a chip shot from the US. Even modest stations could make contacts on several bands. Navassa is an uninhabited wildlife sanctuary administered by the US Fish and Wildlife Service, and they won't let anyone on the island. Fish and Wildlife has already decreed it will be at least 2025 before they allow another operation. Glad I got mine.

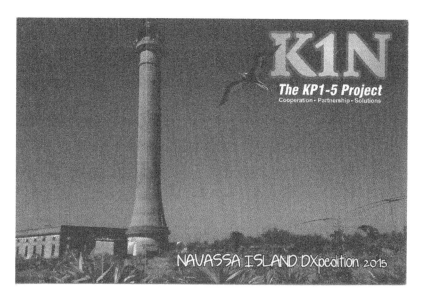

---

[10] The original K4IA, Ed Loller, was on the DXpedition to Navassa in 1969.

VERY RARE DX

Desecheo Island, off Puerto Rico, was activated in 2009 after a 15-year hiatus imposed by the Fish and Wildlife Service. Desecheo was a bombing range from World War II until 1952, so watch your step if you go there.

Navassa and Desecheo are examples of rarity caused by government decree.

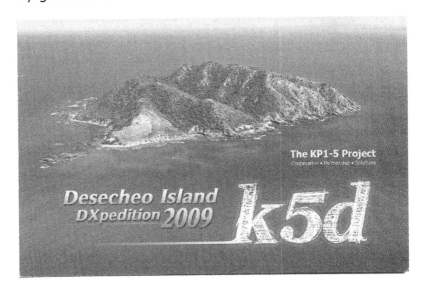

Bouvet and Scarborough Reef are examples of rarity caused by isolation and remoteness. Scarborough Reef is a minuscule group of rocks. The Philippines and China dispute ownership, and China is building up the area in a controversial move to establish an airbase controlling the South China Sea.

Scarborough Reef has only activated three times, the last being BS7H in 2007. Since the area is now disputed territory, that may be the last time for a while. The QSL card shows the operators perched on platforms built over the rocks. Those were brutal operating conditions for them, and the pileups were insane.

VERY RARE DX

Since Scarborough is 8,600 miles from the East Coast, this was anything but a chip shot. I felt extremely fortunate to work them on 20-meter CW. The solar flux index at the time was barely enough to support even a 50% chance, according to my propagation prediction program. Don't live and die by the predictions.

VERY RARE DX

Hone your operating skills to work the Very Rare DX. Learn CW and RTTY and how to work split. Working Very Rare DX requires patience, listening, lost sleep, equipment tweaks, listening some more, and finally, good old-fashioned luck. The better you are, the less luck required. Advanced techniques are in coming chapters.

Take heart. A good DXpedition will work thousands of stations, maybe tens of thousands. All those contacts are not California Kilowatts with monster towers and beams. Even QRP stations make it in the log. Your station can be competitive even if you think it is modest. Finesse overcomes power.

Here's a card from a Micro-lite DXpedition to a Very Rare DX entity.[11] Read what they have to say about a workable signal into any corner of the earth.

"*The Micro-lite Penguin philosophy celebrates simple antennas such as verticals and dipoles. We typically use 100 watts. That is enough to put a workable signal into any corner of the earth.*" Lew Sayre W7EW/W7AT for the Micro-lite Penguins.

---

[11] VP8, South Sandwich Islands, is #21 on the ClubLog most-wanted list.

DX CODE OF CONDUCT

Operating in a DX environment is not the same as casual rag chewing. You don't talk to your boss the same way you converse with a close friend. The expectation is for more formality and precision. An international group of DXers developed the DX Code of Conduct to restore more gentlemanly behavior and inspire a higher ethical standard. The website is Dx-Code.org.

Here is the DX Code of Conduct with my comments:

1. I will listen and listen and then listen again before calling.

Listening is essential. Tuning and listening provide a chance to find the DX first while he is still alone. Once a spotting network posts the DX, you have to compete with everyone (and I do mean *everyone*).

When you hear DX, listen to what is happening before jumping in:
- Do I have the callsign correctly?
- Is the DX working split?
- Where is he listening?
- What are his pattern and cadence?
- Is he working by geographical area or by-the-numbers?
- Is he finished his QSO and ready to receive calls?

Listen first and only transmit after you are sure that you will not be a LID.[12]

Practice developing the skill to answer those questions quickly.

---

[12] A LID is a poor operator. You don't want to act stupidly when the whole world is listening.

DX CODE OF CONDUCT

2. I will only call if I can copy the DX station properly.

Sometimes, the DX operates with less power or antenna than you have. You can barely hear him. What good is it to call when you can't hear the DX come back? That is only adding QRM. If you can barely hear the DX, wait until a momentary shift in propagation brings his signal up to a copiable level. Then, and only then, make a call. Otherwise, wait and come back another time when you can hear better.

If the DX answers and you don't respond, he will try again, maybe only one more time. After that, he moves on, and the whole world knows you missed the call. I have heard this many times as a deaf caller keeps pounding away, trying to shove himself into the log. After a few tries, the DX gives up, and the opportunity is gone.

3. I will not trust the cluster and will be sure of the DX station's callsign before calling.

DX spots can be wrong. A cluster may spot an HA in Hungary as a 5A in Libya. It is only a dit worth of difference on CW. Log him as a 5A thinking you just bagged a new country, got nothing. You'll be wondering years later, "How come that bum never answered my QSL card?"

Sometimes, there is more than one DX station on a given frequency or very close to it. DX tends to congregate on frequencies called the DX window. Two stations in Africa could be on, or close to, the same frequency, but they don't hear each other because they are in the skip zone. They are both running a pileup, and spots could identify either one. Somebody gave a 59, but which station did you work? Listen to verify the callsign independently of the spotting network.

DX CODE OF CONDUCT

4. I will not interfere with the DX station or anyone calling, and I will not tune up on the DX frequency or the QSX (split) spot.

Rude, illegal, and destructive behavior takes lots of forms. Don't tune up on the DX frequency or in the middle of the pileup. People do it all the time. Another example is deliberate interference (DQRM) from a frustrated operator sending dits or a carrier on the DX frequency.

Don't make comments on the DX frequency, such as: "Geesh, the band is open to West Coast, and he's still working Europe." "When is he going to work NA?" "Please move to 20 meters."

There are frequency cops, "self-appointed policeman." They go to work when some poor soul makes a mistake, like forgetting to operate split. They scream "UP UP you #*% LID" on the DX frequency. Don't engage in arguments; ignore them.

A pet-peeve of mine is the CW operator who feels compelled to send "?" every time the DX does not come back to him. If you didn't hear who he called, listen. If the DX doesn't get a response, he will call again. Sending a question mark is nothing but DQRM.

5. I will wait for the DX station to end a contact before calling.

Interrupting is rude. Squeezing a call in when someone takes a breath will not get you in the log.

Improper "tail ending," calling before the DX finishes the contact, is also interrupting. It breaks the cadence, interferes with the exchange, and aggravates the DX operator. Just because the DX says, "73" does not mean he finished. The other station still has a final. Wait for the DX to say his call, "QRZ" or "up," indicating he is ready to receive callers.

DX CODE OF CONDUCT

6. I will always send my full callsign.
Sending a partial callsign requires an extra exchange to get the full callsign, slowing down the pace. Without a full callsign, these are unidentified and, therefore, illegal transmissions. Net check-ins sometimes use partial calls, but I can't imagine why.

I once heard a DX op say, "KWA, you're in the log." What he meant was, "I put 'KWA' in my log just like you said. I am not going to ask for your full call." Of course, without a full callsign in the log, Mr. KWA is out of luck and probably blackballed or put in time-out until tomorrow.

7. I will call and then listen for a reasonable interval. I will not call continuously.

Continuous calling keeps you from listening. You might miss the call back because you are doubling – transmitting when you should be receiving. I have often heard a caller stomp on his own contact because he is calling the DX at the same time the DX is calling him. The DX can't hear the chronic caller because they are both transmitting at the same time. The DX will move on, and the chronic caller will never know what he missed.

Continuous calling also adds to the QRM. Give the DX a chance to respond. If he works someone else instead of you, that is one less person in the pileup. Be thankful that your odds just got better.

8. I will not transmit when the DX operator calls another callsign, not mine.

If the DX called me, do not call on top of my reply. If the DX doesn't hear me because of your interruption, he will call me again. A good DX operator will not answer you, except perhaps to chastise. Your out-of-turn calling slows the exchange and earns the scorn of everyone who hears you try to muscle your way into

DX CODE OF CONDUCT

the log. Get some manners and go to the back of the line where you belong.

I hear this egregious behavior in every pileup. The DX acknowledges a specific station, and dozens of LIDs, whose callsign could not possibly be mistaken for the one called, insist on sending their callsign. It NEVER works. It NEVER results in the DX calling the LID. The DX is listening for a signal report from the station he called. These extraneous callers are QRM, both for the DX and the good operators listening for the called station.

9. *I will not transmit when the DX operator queries a callsign, not like mine.*

If the DX says, "Who's the W4?" everyone else should stand by until he works the W4. Don't try to convince yourself that he must have meant you because you have a W somewhere in your callsign. The DX didn't misunderstand anything, and you sound ridiculous when you call. This rule is ignored far too often and will turn a pileup into chaos if not properly managed.

If he is indeed calling me and misunderstood, he will try again, and then I can decide if he meant me or someone else. Listen, think, and then act.

Some people copy my callsign on CW as K4EA instead of K4IA. What a difference a dit makes. Mistakes happen on voice as well. When they do, I listen for him to call K4EA again. If he does, it means K4EA did not answer the first time, and I can be pretty sure he means me. I've got a high confidence level because 3-out-of-4 letters match up, and I know, from his QRZ.com page, that K4EA is a big VHF operator. I feel comfortable sending "K4IA IA." I add the "IA" or "India Alpha" for emphasis and not a substitute for sending the full call.

DX CODE OF CONDUCT

Once the DX has my callsign straight, then, and only then, do I send the signal report. If I send the report too early, he will think we finished and log me with the wrong callsign. Once the DX is on the hook, don't let him off until you are double-plus sure he got your call correctly.

10. I will not transmit when the DX operator requests geographic areas other than mine.

There are several reasons a DX operator may restrict callers by geographic areas. One, it thins the herd. Second, there may be a propagation window to that part of the world, and the DX wants to give those folks a chance. If you hear "EU," don't call if you're not in Europe.

The DX may also thin the herd "by-the-numbers," taking callsigns containing a specific number for a few minutes and then moving on to the next number. A good DX operator will ignore out-of-turn callers and blackball persistently bad operators. Be patient, listen to the pattern. Take a break and come back when you think he will get to your number.

Good DX operators, and certainly the DXpeditions, try very hard to give everyone a chance. You don't know what they hear on their end of the propagation path. Trust that they are going to get to you, and whatever instructions they give are for the benefit of all.

11. When the DX operator calls me, I will not repeat my callsign unless I think he got it incorrectly.

Say "59," and the DX will log the callsign he has. If you repeat your callsign, he will assume he got it wrong the first time and that you are trying to correct the record. He will ask for clarification, and that confuses and slows down the whole process.

DX CODE OF CONDUCT

VERY IMPORTANT > If the DX got your callsign wrong, or you are not double-plus sure, give only your correct callsign in return. Make sure he has it correctly before you give the signal report. Otherwise, he will hear your report, assume all was well and go on to the next caller. Then, the wrong callsign will end up in the log, and you'll be wondering, "Did he catch my correction?" He probably didn't. If the DX catches a correction, he should repeat the correct callsign. Don't take a chance – callsign first, then report.

If you suspect a case of mistaken identity, take a break and call again in a few minutes. Use this option carefully as it is considered very bad form to make duplicate contacts. Dupes deny others the chance to make even one contact.

12. I will be thankful if and when I do make a contact.

Gratitude is a matter of attitude. Some of these DX operators, particularly those on DXpeditions, go to great personal sacrifice and expense so you can put a QSL card on the wall. Thank them. Don't make disparaging remarks or criticisms. You have no idea the hardships DX operators endure.

Please contribute to the DXpedition's expenses, either on their website or throw some money in the envelope with your QSL card. Don't disrespect a foreign ham who asks for a few Green Stamps[13] to cover his cost to send that coveted QSL card.

---

[13] Dollar bills.

DX CODE OF CONDUCT

13. I will respect my fellow hams and conduct myself so as to earn their respect.

Mistakes happen. We have all called on the wrong VFO, forgot to set our transceiver to split, called at the wrong time, and committed all manner of faux pas. Fix the problem and move on. No need to make apologies on the DX's frequency. If a buddy hears and decides to rib you about it, offer to buy him a beer and have a laugh – just not on the DX frequency.

The whole world can hear you and your callsign. Get a reputation as a bully, cheater, QRMer, or a policeman, and you won't get over it soon. DX can be competitive, but we must maintain balance if it is to remain fun.

Don't be an elephant stomping over everyone attempting to make the contact.

DX ETIQUETTE, DECORUM AND PROPRIETY

Etiquette, decorum, and propriety imply observance of the formal requirements governing behavior in polite society. Etiquette refers to conventional forms and usages: the rules of etiquette. Decorum suggests dignity and a sense of what is becoming or appropriate for a person of good breeding: a fine sense of decorum. Propriety implies established conventions of morals and good taste: She never fails to observe the proprieties.[14]

I will lump them all together into "do's and don'ts" of operating in pileups. Listening to a DX pileup, you will experience every one of the "don'ts" listed below. It is shameful.

Here are my DX Proverbs derived from the Code of Conduct:

Do listen before calling.
Don't call unless you can copy the DX, know where he is listening, and know his cadence.

Do listen more than you transmit.
Don't use "sharp elbows" by calling continuously and stomping over the competition (and yourself).
Don't call when the DX calls someone else.

Do listen or go to the Internet with questions about the DX.
Don't interrupt on the DX frequency to ask stupid questions like, "Who's the DX?"
Don't ask, "What's the QSL route?" in a DX pileup.
Don't ask the DX in a pileup to check if you are in the log for another contact.

---

[14] Dictionary.com

ETIQUETTE, DECORUM AND PROPRIETY

Do make QSOs by yourself without help.
Don't ask for help and don't "help" others. The rules say a contact should be mutual and not assisted by someone else filling in the information. "He says you're 59" is not a valid exchange.
Don't ask a busy pileup to stand by while your QRP friend makes a call. That might be OK with a casual DX QSO but not in a busy pileup.

Do wait before calling again.
Don't QRM your QSO by calling over his reply.

Do wait for your turn.
Don't tailend.
Don't call when the DX calls someone else.
Repeat: Don't call when the DX calls someone else.
Don't send "?" or ask "QRZ?" when the DX calls someone else.

Do call with your full callsign.
Don't use partial calls.

Do use standard phonetics.
Don't use cutesy phonetics.

Do work the DX once per band and mode and check any online log to confirm your contact.
Don't pile on with dupes.

Do pay attention to the operator's instructions.
Don't call if he is working another area or number.

Do honor the DX's requests for ATNO only (All-Time New Ones only).[15]
Don't deny those who have never worked the entity a chance.

---

[15] The DX is asking for only those stations who have never worked this entity on any band or mode

ETIQUETTE, DECORUM AND PROPRIETY

Do tune up into a dummy load or clear frequency.
Don't tune up on the DX frequency or in the middle of the pileup.

Do exercise discretion in politely correcting others.
Don't be a frequency cop. Yelling, "Up, you idiot," doesn't help and only contributes to QRM and chaos. You will probably get an "FU" in response.

Do ignore smart-alecks, jokesters, QRMers, and malcontents.
Don't answer them. It only encourages more bad behavior. Leave them alone, and they will get bored and move on.

Do check your transceiver for split operation before calling.
Don't call on the DX frequency.

Do listen within the frequencies designated as DX windows. DX windows are part of a voluntary band plan setting aside specific frequencies for DX.
Don't hold domestic rag chews in the DX windows.

Do double-check the callsign before spotting DX on a network.
Don't spot a callsign if you're not sure.
Don't "over-spot" easy DX or duplicates.

Do verify the callsign independently before putting the contact in your log.
Don't rely on the spotting network for the callsign.

Do work the DX and be happy about it.
Don't try to move a pileup to another band or mode. You're not in charge. Asking takes time that denies others a chance. Who are you that the DX should stop what he is doing, ignore those waiting and go to another frequency to satisfy your one pitiful request?

ETIQUETTE, DECORUM AND PROPRIETY

Do maintain a tolerant worldview. Talk about ham radio and not religion or politics.
Don't ask questions like, "When are you going to get rid of that bum dictator?"

Do use the spotting network to spot relevant and unique DX.
Don't pile on the spots. If another station in 4-land has just spotted the DX, there is no reason for me to do it again. Doing so clogs up the system.

Do use the spotting comments section for information useful to others. "Lonely" tells the world there is no big pileup.
Don't use it for personal conversations or mean comments. "Deaf" "Poor op."
Don't be a hot-dog posting your triumphs with comments like "Finaallllly!" "BINGO!" "First call!!!"

Do identify with your callsign.
Don't make anonymous rude remarks. If you are not willing to put your callsign on something, you shouldn't say it.
Don't make unidentified illegal transmissions.

Do listen, listen, and listen some more.
Don't call when the DX calls someone else.
Repeat: Don't call when the DX calls someone else.
Repeat: Don't call when the DX calls someone else.
Repeat: Don't call when the DX calls someone else.

OPERATING TIPS AND STRATEGIES

Tricks and skills set the experts apart. A few operators get through by the brute force of their station's equipment. The rest of us must rely on finesse. Finesse can grab the bone from under the big dog's nose. Let's talk about some skills you should develop and apply.

TRAIN YOUR EARS

You've heard over and again to listen, listen, and listen some more. How do you listen well? First, train your ears.

My first receiver was a military surplus ARC5 modified to spread the 40-meter CW band across the dial. I didn't need to tune the dial because the receiver was so broad, it seemed I could hear the whole band at one time. There were more than a dozen CW signals on different frequencies, heard at different pitches, at once. I learned an essential skill – to concentrate on one signal and mentally block out the rest.

For CW, learn to concentrate on one CW pitch and that tone only. Ignore everything else. Don't be distracted by other signals, even if they are stronger. You will not copy anyone correctly if you allow your concentration to jump around,

It is a little harder on SSB. Off-frequency SSB is very distracting even if it cannot be understood. Get close, concentrate on one signal, and mentally tune out the rest.

OPERATING TIPS AND STRATEGIES

ZERO BEAT

Zero-beating is setting your transmit frequency to the same frequency as the other station.

On CW, match the tone heard in the receiver with the side-tone of your transmitter. To implement this skill on CW, pick a comfortable sidetone pitch and learn to recognize it. For example, I like 650 Hz. If I adjust the receiver to give a tone of 650 Hz, I will be zero-beat. There is often a "spot" button on the transceiver that will inject a fixed tone. Tune the VFO to match that tone with that of the received signal.

I tend to tune an SSB signal too high in pitch. That means he will hear me too low when I call. It is harder to understand low. I have learned to compensate.

LISTENING

Listening can take several forms: audible listening, spotting networks, and panadapters.[16] Spotting networks and panadapters are The Easy Way, and we'll cover them in other chapters.

Listening is an important skill to cultivate. Try to find the DX before anyone else, especially before the spotting networks pick him up. Once he's spotted, you become one of many calling instead of the only one calling.

The contesters have an expression called BIC. It stands for "butt in chair." To make contacts you must be willing to spend time with your butt in the chair.

---

[16] A panadapter displays signals on a screen. You can look at the display and see if there are any signals on the band without having to tune around and listen.

OPERATING TIPS AND STRATEGIES

Listening can be very productive because:
- You hear when and where the band is opening
- You may hear a DX station calling CQ
- You may discover a rare DX station before he is spotted on one of the networks, providing a chance to work him before the horde descends
- It's fun.

Tune slowly and listen to every signal until you identify the participants. In time, you will develop a sixth sense to differentiate DX from domestic. For example, auroral flutter is a dead giveaway for over-the-pole propagation to Asiatic Russia, India, and the Far East. Spoken accents may reveal the operator's native language. Raspy-sounding CW may indicate a multi-path DX station or a power supply issue. Power isn't too clean in many parts of the world, or the station may be using an old power supply with leaky capacitors.

I prefer to tune in a direction such that the received station's tone starts high and descends. That way, I can hear the signal coming. The high-pitched tone cuts through. When tuning from low-pitch to high, you can't hear until a low rumble ascends to something understandable.

Notice I did not say to tune from high frequency to low but from a high tone to low. Frequency-wise, the direction depends on which sideband or which side of the CW you are listening. Try this, and it will make more sense.

OPERATING TIPS AND STRATEGIES

USE PASSBAND TUNING

A 2.1 kHz filter is narrow for conversational SSB. Crowded conditions might call for a 1.8 kHz filter. Those are the standard narrow crystal filter widths.

A transceiver with DSP[17] filtering is not limited to one or two bandwidths provided by crystal filters. You can crank the bandwidth down to whatever you want. By default, the radio puts the passband in the middle of the filter. To copy speech, concentrate the receiver on the portion of the signal containing the higher audio frequencies. That is where the passband tuning comes in.

I was operating in the Virginia QSO party on 40 meter SSB when the QRM was so bad that I could hardly copy anybody. I found that I could copy if I narrowed the DSP below 1.8 kHz and slewed the passband.

The same principle can work with a CW signal if there is QRM nearby. Try working the passband tuning to get away from the offending QRM.

REDUCE YOUR RF GAIN

Reducing gain may seem counter intuitive. After all, don't you want to hear weak signals? Modern receivers are extremely sensitive. Sometimes, they are too sensitive. Being able to copy is all about the signal to noise ratio. You want to hear the signal, not the noise. Turning up the RF gain may increase the signal, but it also increases the noise level. The signal is still buried in noise.

Turning up the RF gain can be detrimental. Receivers have a limited dynamic range. Exceeding the dynamic

---

[17] Digital Signal Processing is filtering using math. The filter shape and width are adjustable.

OPERATING TIPS AND STRATEGIES

range introduces distortion by-products. It can also overwhelm the AGC (automatic gain control) circuits, causing them to mute reception.

If the noise level is already high, as it usually is on the lower HF bands, reduce the RF gain or switch in the attenuator to reduce the noise level. Turn up the RF gain just to the point you start to hear the background noise and no further. This setting is less fatiguing for the ears and preserves headroom for the receiver's dynamic range limitations.

Never use the pre-amplifier except on high-frequency bands (15 meters and up) when the noise is low. I remember watching over the shoulder of a 75-meter phone position one Field Day. The operator had the pre-amplifier on, and the RF gain cranked up to 11. He was complaining that the noise made it impossible to copy. I turned off the pre-amp, reduced the RF gain, and dialed in some attenuation. Amazingly, the band cleared up, and signals were readable again.

OPERATING SPLIT

Zero beat when operating simplex, but the strategy changes when chasing a pileup in split mode. Working split requires listening to two different frequencies, the DX, and the callers. I'll call the DX the "fox" and the callers "hounds." The DX fox stays in one place, but the hounds chase around to find the frequency where the fox is listening. Listen to the pack of hounds to hear who is successful. When you hear "59," you know which fox was triumphant, and that is where the DX was listening.

Ideally, you have a transceiver that will let you hear both the DX fox and the baying pack of hounds even though they are on different frequencies. Radios implement this technique differently. The ideal layout

OPERATING TIPS AND STRATEGIES

is separate receivers outputting stereo audio with one receiver in each ear.

The less ideal is more like a Receiver Incremental Tuning system that outputs both signals into a mono audio output. Others require a button push to switch the receiver from VFOA to VFOB. You only hear one at a time, and the sound is mono.

In split mode, the transceiver transmits on VFOB, where the hounds are.

Put a narrow filter on VFOA to hear the fox. That makes him easier to copy because it cuts down on the noise and interference outside the passband. Narrower filters improve the signal to noise ratio and always help reception.

Set a wide filter on VFOB to hear the hounds. The hounds spread out and won't call on precisely the same frequency. Hearing a range of frequencies allows you to identify who is answering the fox and tune VFOB to him. In split mode, concentrate on hearing the 59 or 5NN signal report from the lucky hound. Then you know where the DX fox was listening. What do you do with that information?

Strategy number 1 is to match the last hound's frequency exactly. Zero-beat the hound, hoping the DX is still listening on the same frequency. Zero-beat often works in light pileups. If you've been listening, you will have noticed if the fox keeps working hounds in the same place.

Strategy number 2 is to call slightly off the last hound's frequency. Figure that everyone else will jump on that frequency, so the QRM from the rest of the hounds will drown you out. Be somewhere else. Listen and discern a pattern. Is the fox listening in one place, moving up, moving down, or jumping all

OPERATING TIPS AND STRATEGIES

around? Get one step ahead by calling on the frequency where you expect him to move, not where he has just been.

If you get lost and can't hear the successful hounds, but know the fox is moving up, go up a bit and call there. Wait for the fox to come to you. If he is moving upwards, he will surely tune across your signal.

A panadapter will show the bouncing trace of a hounds' answer. When the DX calls a station, only that station should answer. His trace should be the only one on the screen (except for the LIDS calling out-of-turn). The panadapter might also show a hole where no one is calling. Drop in the hole and give it a shot. The fox is tuning, looking for clear signals, and you may stand out enough to be heard.

TIMING

Timing is everything in life, and it is essential when calling DX. DX Code of Conduct #5 cautioned us against improper tail ending. What differentiates proper and improper tail ending?

Improper tail ending, calling before the DX finishes, is interrupting. The DX fox may have stopped transmitting, but he is still listening to the hound. A tail-ender stomps on the hound, frustrating the fox and upsetting the QSO's timing. Now, the DX has to ask for a repeat, "Sorry old man, someone called and stepped all over you. Can you repeat?"

Wait until the fox says "QRZ" or gives some other indication the conversation is finished. Proper tail ending is waiting to be sure the previous conversation is over before you call.

OPERATING TIPS AND STRATEGIES

You also need to know when to pause. Look back at Code of Conduct #7. If the hounds are all barking simultaneously, calling and calling, the QRM level builds to a crescendo. Wait for a clear moment, and you may be the only one calling. I've had that work for me many times. The big dogs were baying incessantly, but the DX fox didn't come back to anybody, so it got real quiet. I quickly slipped my call in and was immediately recognized. The fox sounded surprised, "Wow, K4IA, that was good timing. You're not real strong, but you were all alone." I felt like a champ. Finesse beats power. I had let the pack die down and threw my call into dead air. The fox was overwhelmed by so many stations calling. He couldn't copy anybody until he heard me all alone.

Sometimes the fox will pick out callers right away. In that case, waiting gets you nowhere. Listen and learn the fox's operating style. If he responds to quick calls, give him one. But if there is silence, jump in.

However you call, wait for an answer. Learn the cadence. I have often heard the fox come back, but the hound is already calling again, so he doesn't hear the fox. The fox will try one more time, and, again, the hound is calling when he should be listening — the two talk past each other, and the fox moves on to another call. The hound misses the contact and never knows how close he was.

PHONETICS

On voice modes, standard phonetics help cut through noise and QRM. They are essential when dealing with speakers of a different language. If you say "A as in date" to a Hispanic, he sees the letter "E," and saying "E" to him is the letter "I." No wonder callsigns get busted.[18] Using standard phonetics makes it easier for

---

[18] A "busted" callsign is one that was copied incorrectly

OPERATING TIPS AND STRATEGIES

other-language speakers to understand. Notice I said, "standard phonetics." Cutesy or non-standard phonetics confuse.

Here is the phonetic list adopted by the International Telecommunication Union:

A--Alpha	**J**--Juliett	**S**--Sierra
B--Bravo	**K**--Kilo	**T**--Tango
C--Charlie	**L**--Lima	**U**--Uniform
D--Delta	**M**--Mike	**V**--Victor
E--Echo	**N**--November	**W**--Whiskey
F--Foxtrot	**O**--Oscar	**X**--X-ray
G--Golf	**P**--Papa	**Y**--Yankee
H--Hotel	**Q**--Quebec	**Z**--Zulu
I--India	**R**--Romeo	

Be consistent within your callsign. If you are Kilo Alpha 4 India Alpha, don't say Kilo America 4 India Alpha. "America" and "Alpha" are different, and your DX may wonder what he heard.

However, if the DX is having trouble copying "Alpha," try saying "America" in both places and hope that makes more sense to him. Just be consistent.

When calling, be clear and crisp. Don't mumble, don't shout, and don't stretch out your callsign. Keeeeellllooo is not easier to understand than Kilo. It can sound like several letters – Kilo, Hotel or Tango depending on which part of the elongated word was heard. Keee can sound like Kilo. Ell can sound like Hotel. Oooo can sound like Tango.

If the DX is having trouble with your number, it may help to count for him. "One, two, three, four. Kilo Four India Alpha."

OPERATING TIPS AND STRATEGIES

CW STRATEGIES

We've already talked about zero-beating and matching the split. What else can you do?

When calling DX, other than casual DX, send your callsign one time. Don't send his callsign at all. Just answer him. A single call is becoming the standard operating style on SSB as well. The DX will assume you must be calling him if you are zero-beat on his frequency or in the pileup. Measure if the DX is experienced or not by his style of calling CQ and his speed. If he seems unsure, try the standard "his call de your call, your call," except in a pileup.

No matter how you're calling, don't use a "K" to signal the end of a CW transmission. Leave the "K" off; silence is enough to let the DX know you are done sending. If you send "K4IA K," the DX will think your call is K4IAK. If you already have three letters, the DX will be confused. "WA4TUF K." What did I miss? Was it WA4XFK?

Match the fox's CW speed. Usually, these guys are going at a fast clip – maybe faster than your comfort level. Learn to recognize your callsign, even at high speeds, and match his speed when sending yours. Send too slowly, and he may never come back to you or will already be working someone else by the time you finish sending. Calling too fast may exceed the fox's comprehension, particularly if your's is a long and complicated callsign.

Another suggestion is to match the speed of the stations whom the DX answers. The DX may have set his keyer to send faster than he likes to copy. He does this to speed up the exchange, but he responds to stations calling at his comfort level. Listen for a pattern and match it.

OPERATING TIPS AND STRATEGIES

Using QSK, a rapid changeover from transmitting to receiving will help you hear the DX come back between characters as you send.

GET A NEW CALLSIGN

Consider getting a vanity callsign. Pick one that is not a mouthful or confusing on CW. Your initials may not be the best choice. I attribute a lot of my success to my call, K4IA. It sure is more manageable than KG4CVN; assigned me when I relicensed in 1999.

Pity this poor team of operators on Tromelin, 280 miles east of Madagascar. Tromelin is a mile long and a half-mile wide sandbar in the Indian Ocean. The four-man team made 50,000 contacts. Can you imagine having to say or key "FR/F6KDF/T" 50,000 times?

ANATOMY OF A QSO

Now you've learned the rules. It's time to look at an actual DX QSO. First, we'll consider how Larry the LID might go about it.

Larry has a big boat-anchor station with lots of power and one of those Windom antennas from Carolina with 13dB gain in all directions (or so it is claimed). Larry likes to be loud and likes getting his way, so patience is not one of his virtues.

It is getting about time for Larry and his buddies to hold their daily "net" (gabfest). He can't wait to hear about Moe's gall bladder surgery and Curley's problems with the taxman. He's anxious to share his story of setting the counter-girl at BurgerDoodle straight. She had messed up his double-bacon-diablo-big-bull order. "Yeah, man, I had her in tears. I can't wait to tell the guys all about it."

Larry turns on his radio and is much chagrined to find there is someone right on HIS frequency. "Damn, how am I going to shoo him away? Let's start by checking my amplifier; surely, it needs tuning up." After a lengthy key-down, Larry switches back to receive, and the station is still there. Larry doesn't realize it is a DX station listening up, so the DX isn't fazed by Larry's tuning up on the transmit frequency.

The rest of the world, listening on the transmit frequency, is going crazy. Larry can't understand all the catcalls and abuse directed his way. "Tune up somewhere else, you LID!" "*&^% idiot!" "Wrong VFO Bozo." "He's listening up, up, up, up, up, up!" "Spleeeet Spleeeet" "DX is on the frequency!" "QSY!" Larry isn't fazed a bit and calls out, "Who's the DX?" That sets off another round of abuse. Larry thinks, "Gee whiz, what a bunch of crabs. I'll check on the cluster and see who it is."

ANATOMY OF A QSO

Larry fires up his computer and loads his favorite DX cluster. He hasn't used it in a while, so it takes some time to remember how to get the system working. Larry is a bear of little brain and can only do one thing at a time. He turns down the radio to help his concentration and, as a result, doesn't hear the DX identify. Finally, Larry succeeds and finds a spot listed for a GM callsign pretty close to the frequency. "That must be him. Scotland ain't no special DX. I'm going to find him and tell him to move."

It never occurs to Larry that maybe he should listen to identify his "intruder." Larry doesn't bother to listen. That is too much trouble. He doesn't have dual receivers or any way to hear both sides of the conversation. Larry cranks his transmit incremental tuning up a few and starts calling, and calling, and calling. "GM5KIM, GM5KIM, Greasy Microphone Five Keeping It Messy, please copy W4LID,[19] W4LID, W4LID, Wanted For Living In Dreamland. DOOOOOUBLEUUUUUU FOOOOOOR ELLLLLLLLL IIIIIIIIII DEEEEEE. 59 59 Over Over" No reply, so he calls again and, again and again, repeating "GM5KIM, GM5KIM, Greasy Microphone Five Keeping It Messy, please copy W4LID, W4LID, W4LID, Wanted For Living In Dreamland. DOOOOOUBLEUUUUUU FOOOOOOR ELLLLLLLLL IIIIIIIIII DEEEEEE. 59 59 Over Over"

Somewhere in there, the DX has identified several times but, since Larry is transmitting, he doesn't hear. Now Larry is getting mad and posts a note on the cluster accusing GM5KIM of being an alligator – all mouth and no ears. "This guy must have mush for brains, or he is deaf." Larry keeps pounding away, twitching his transmit incremental tuning.

---

[19] I worked the real W4LID, who said he didn't mind me using his callsign for this example. He's a good operator and a good sport.

ANATOMY OF A QSO

Eventually, a station comes on the DX frequency, "W4LID, he is calling you." Larry thinks, "Well, I don't hear him. The dumbass must have turned off his amplifier." "Hey, you're on our frequency. You need to move right away. We've been using this spot for years."

Then Larry hears, "Sorry Old Man, this is HM5KIM from North Korea." Then, with dripping sarcasm, "I certainly wouldn't want to inconvenience you. HM5KIM is going QRT for a while. Enjoy your frequency."

Now the entire world is howling at Larry. "You chased him off you piece of %^*." "Are you kidding me?" "I'm on my way over to shoot you, you &^%*."

Larry just smiles to himself. "I guess I told him." Moe and Curley will be on in a few minutes. Wait until I tell them how I saved our frequency."

This scenario is not far-fetched at all. "G" is right next to "H" on the keyboard, and someone could easily fat-finger the spot entry. Not that Larry would care. The only far-fetched part of the story is assuming an operator in North Korea.

Let's contrast Larry the LID's experience with one I had just a few days ago. I was listening for the DXpedition to Palmyra[20] on 15 meter SSB. My prediction program told me the band should be open. I knew they had been operating around 21.200 MHz because I checked the spotting history, so that is where I am. I have my antenna pointed to due west, and I've tuned my amplifier. I set my transceiver in split mode. I am at the right place at the right time and ready for action.

---

[20] Palmyra Atoll is due south of Hawaii, about half-way between California and Australia.

ANATOMY OF A QSO

I hear nothing and see nothing on my panadapter. I don't even see evidence of a pileup on the panadapter. Usually, you see a flurry of activity that signals a pileup even when you can't hear the DX. I've worked Palmyra before, so it is not a big deal, but I enjoy the sport, so I always chase the rare ones.

I see a reasonably strong signal on my panadapter up higher. I'm bored, so check it out. I find a GM in Scotland. I've worked Bob before, and he's got a great signal tonight. Scotland is no big deal, but I don't have it on 15-meters SSB, and Bob uses Logbook Of The World,[21] so I decide to go after him.

I listen for a bit and learn that he is working simplex. He's got a small crowd but not enough to go split. Bob is handing out the QSO trinity at a rather leisurely pace, but there is no extraneous conversation. It's all business, and he is good at it.

I swing the beam around to the Northeast. His signal comes up; now, he is 20 over. This should be easy. Bob is picking stations out and answering fairly quickly, so I wait until he says "QRZ?" and throw out my callsign. He comes back to someone else. Rats. Try again. No luck. Try again. No luck.

Now, I am wondering if I have bad breath. Then I think, "Wait. I didn't take the radio out of split mode. I am calling on the wrong frequency." You probably think Larry the LID was my Elmer. It was an easy mistake to make. It was unintentional. I won't characterize that as a LID mistake, but it was embarrassingly close.

Now, I am in simplex. The crowd is thickening, and I notice Bob is slowing down. He's not answering the

---

[21] An online QSL service. See the chapter "Confirming Your Contacts."

ANATOMY OF A QSO

first caller. I wait until he finishes, and I slowly count to three before throwing out my callsign. It seems like an eternity, and I am certain someone else will get in before me. "Kilo Four India Alpha." After a pause, he answers another station. He's taking longer. I need to slow down.

This time, I count to five before calling. Bob comes back to me with a 59+ signal report and "Nice to hear from you again, Buck." Obviously, he has a computer logging program that pulled up our previous contacts. No one would remember my name otherwise. I reply "Very good Bob. You're 20 over here in Virginia. Good luck and 73." That's it. He could have volunteered his operating conditions and asked me about mine. He could have asked me if I worked the Palmyra folks. He could have told me about his weather. But he didn't, so it was not up to me to drag the QSO out any longer than he wished. It was Bob's frequency, and he set the pace. He was in DX mode, trying to give many people a contact with Scotland.

This isn't Bob's card, but it is definitely Scotland.

PROPAGATION

Propagation changes depending on many variables, and there are plenty of sources for up-to-date predictions.

Just remember, they are predictions. Models are designed to predict the odds of an open path. What odds do you need? If you are shooting for a 99% certainty, the predictions won't show many openings. Will you accept a 75%, 50%, or 10% chance? Are the odds better on a different band? You are not betting the farm here. There is no "lose." Long shots sometimes win. Look at the predictions as suggestions, not iron-clad rules.

You might accuse me of being unscientific, and I accept that. My point is that you should not despair just because the predictions say there is little chance. Even a little chance can pay off.

PREDICTION PROGRAMS

There are several sources for propagation predictions. The DxLab Suite[22] includes a module called PropView, which will generate a Maximum Usable Frequency chart at different confidence levels and show the possible band openings to the DX.

You can also input your location and station parameters along with those for the DX in Voacap.com/prediction to generate a prediction chart for various bands and times. Checking the path to Palmyra (KH5) today, it looks like I need to be on 15 meters from 1800 – 2300 UTC for a 100% chance of hearing and being heard. That is certainly good to know and a great place to start. Here's a screenshot

---

[22] More about DxLab in the chapter on logging programs.

PROPAGATION

that doesn't do justice to the web page but will give an idea of the power of this tool.

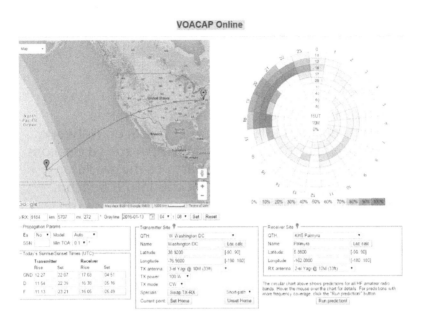

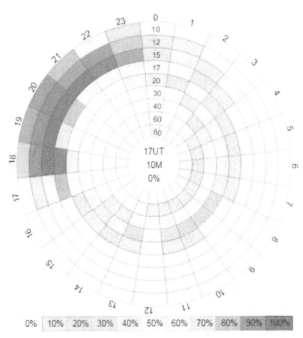

PROPAGATION

The bottom chart is a close-up of the prediction. The dark ring is 100%. Each ring is a band, and the outer numbers are time in UTC. Prediction programs are a powerful tool, but they can't tell the whole story. Predictions are just predictions.

BEACONS

To know what is happening instead of what is predicted, check the NCDXF/IARU[23] Beacon Network. These two volunteer organizations operate 18 beacons that broadcast from different parts of the world on a 10-second rotating schedule. Tune into a beacon frequency, and you will hear some, but not all the beacons as they switch from location to location every 10 seconds. It takes 3 minutes to cycle around the world. By the time a cycle finishes, you will have a very good idea of the band's condition.

Location of beacon stations.

Each station identifies in Morse code at 100 watts using a simple vertical antenna. Then, it sends a dash at 100 watts, 10 watts, 1 watt, and 100 milliwatts (1/10 watt). If you can hear the beacon, you know the band is open to that part of the world. You will be amazed at how little difference there is in signal strength as the beacon sends with less power. If you

---

[23] Northern California DX Foundation / International Amateur Radio Union.

PROPAGATION

can hear them at 100 watts, you will almost always be able to hear at the lower powers as well.

Beacon frequencies are 14.100, 18.110, 21.150, 24.930, and 28.200 MHz. Don't transmit there, or you will cover the beacons and incur the wrath of Khan. If you don't know enough Morse code to decipher what you hear, NCDXF.org/beacon/beacontools has apps to show you who is transmitting.

REVERSE BEACON NETWORK

A variation is the Reverse Beacon Network. Receiving stations all over the world feed information about the CW they hear into the system. You don't have to know CW to see the reports.

Call CQ on CW, and your signal will be picked up and reported along with the signal strength. The Reverse Beacon Network website, ReverseBeacon.net, tells how well other parts of the world heard you.

Participate as a receiver by downloading their application and running it on your computer in conjunction with CWSkimmer, a CW decoder.

GENERAL CONSIDERATIONS

Concentrate on the higher bands during times of good sunspots. Good times don't last forever. When the sun isn't cooperating, fill in your totals and 80/40/30.

Propagation's general rules are well-known. The lower bands are predictable. 80 and 40 meters open long at sunset and close down after sunrise. Magic happens along the gray line – the terminator where night turns to day and vice versa. Get on the air at sunrise or sunset, and there is a good chance you will be able to contact DX stations experiencing sunrise or sunset around the same time. The sunset contact with the

PROPAGATION

Philippines I mentioned earlier was an example of grey-line. I was at sunset, and it was close to sunrise on his side of the world.

A problem with 80 and 40 is that noise from thunderstorms even thousands of miles away, and, in the summer, that can be a real distraction. You also suffer from more man-made noise.

The higher bands are quieter but less predictable. They can be open day or night. It is not unusual to listen to a dead band and hear a DX station pop up out of nowhere. For example, openings from the East Coast to the South Pacific often occur on 20 meters and above around midnight.

CALL CQ

"CQ DX" into a dead band might bring a very surprising contact. If it doesn't, what did you lose? No one is going to laugh at you for calling CQ. And if the band is dead, no one will hear you anyway.

Contesters have a saying, "ABC." It means "Always Be CQing." I maintain that propagation is always open somewhere, but if no one calls CQ, you hear nothing and assume the band is dead. My experience tells me to be brave and call CQ.

Rather than sitting around listening to static, call CQ a few times and see what happens. Better yet, set up a keyer to call CQ for you. A voice keyer stores and plays back a recorded voice message, and a CW keyer does the same for CW. Either can be configured to auto-repeat after an interval, so it keeps going without intervention. It never tires, and never gets bored. Mash the button, sit back, and wait for an answer. (Just don't walk away and leave it on!)

PROPAGATION

Be like a fisherman chumming the water, waiting to see what comes up to take the bait. Read the paper, balance your checkbook, or watch the ball game while the machine does the work. That's The Easy Way.

I remember operating 20 meters on Field Day late at night when 20 meters was "officially" closed. There was no one on the band, so the keyer was set to call CQ while I nodded off. I was jolted awake by a station in the Marshall Islands coming back. He was 7,100 miles away in the South Pacific. Dead band?

I was mobile one Saturday morning, driving around town, when I decided to call CQ on a "dead" 17 meters. My rig was 100 watts SSB and a single-band vertical "stick" on the car trunk. That is far from a big-gun station. A call came back so loud and clear it was as if he was in the seat next to me – only John was in Indonesia, 10,200 miles away. Indonesia is in the Rare DX category. After the contact, I wasn't sure it was real. Then, John sent me a beautiful handmade QSL card – direct, no less. It is one of my favorites. Dead band?

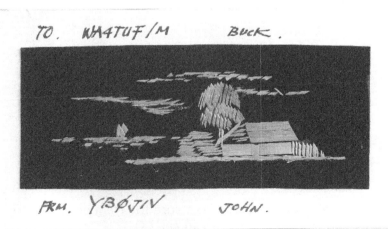

WA4TUF was my original callsign, resurrected under the vanity program before I became K4IA. /M means I was mobile.

PROPAGATION

Reading the solar flux numbers and playing the conventional wisdom, you might conclude certain bands should be dead, but, on a contest weekend, you'll hear many stations. Dead band?

Calling CQ into a "dead" band is a great way to snag DX. What else are you doing? Disturb some electrons. I guarantee the electrons won't complain. DX stations might call confident they will not raise an enormous ruckus with a pileup they don't want to face. If no one answers your CQ, so what?

I shouldn't need to say, listen and ask if the frequency is in use before calling CQ. Even if the band sounds dead, the frequency may be in use, so ask. Send "QRL?" on CW. If the frequency is in use, you will hear "C" in response.

Don't monkey around. Make sure the frequency is clear before you call CQ.

MODES

The mode is the transmission type. You can get by with just SSB, and may even reach Honor Roll, but it is a lot easier if you are flexible. Be ready to operate any mode the DX is using. The discussion below is not meant to be an exhaustive explanation of the various modes; instead, how each mode relates to working DX.

SINGLE SIDEBAND (SSB)

The biggest impediment for beginning DXers operating SSB is "mic fright." If you only operate on the local repeater or have never ventured into the unknown world beyond the usual gang of friends on HF, you might be reluctant to call a DX station. Get over it. The guy on the other end of Casual DX might be as nervous as you are. Practice on him. Get experience working Europeans. They are plentiful and very friendly.

Even for rarer stations, be encouraged! The real pros understand and will try very hard to make the contact. An experienced DXer has heard it all and will help through the exchange. Do your part by listening, playing by the rules, and following instructions. Re-read the chapter on DX etiquette.

The problems with SSB are:
1. It is a wide mode (3 kHz) requiring extra bandwidth. Only a few stations can fit in the listener's spectrum, so there is more QRM;
2. Voice can be hard to understand through interference;
3. Voice takes more power to beat the signal to noise ratio; and
4. Voice is easily misunderstood.

MODES

Pay particular attention to audio quality and equipment adjustments when operating SSB DX. DX settings are different from rag chew settings.

A broad boomy microphone is not best for DX. The majority of speech comprehension is in the relatively narrow band of 400-3,000 Hz. Lower audio frequencies consume most of the power but contain little of the information. Roll off the lower frequencies and increase the upper-middle frequencies with speech processing or choice of microphone.

"Broadcast" quality is not the goal. Concentrate power in the spectrum essential for understanding. Think of the sound from a hand-held bullhorn. It is annoyingly harsh but cuts through.

Adjust the mic gain for the proper drive. Refer to the equipment manual for details on your transmitter. It may have a built-in speech processor to tailor the audio.

After adjusting mic gain, add some compression, but not too much. The audio should not "ring" or pick up background noises. Talk close to the microphone. Being too far away from the microphone requires increasing the mic gain, which picks up more background distractions such as fans, dogs barking, or televisions — no need to add to the existing noise level.

Enlist a friend to make critical comments on your signal with an emphasis on readability, not fidelity. Save the settings and dial them in when you want to work DX.

MODES

CW (MORSE CODE)

There is a chapter on learning Morse code, but for now, we'll cover Morse as it relates to working DX.

CW, which stands for continuous wave or carrier wave, is the most basic of digital modes. The carrier is turned on and off, making the dots and dashes we read as Morse code. You can use a straight key, electronic keyer, or computer keyboard to send. Programming your callsign and 5NN into a keyer is easy. That may be all that is needed.

Receiving is more complicated. You can follow the minimalist CW suggestion mentioned earlier. That is to recognize your call and be familiar enough with the alphabet to decipher the DX callsign concentrating on one letter at a time. This technique doesn't require knowledge of conversational CW but will put contacts in the log.

There are computer programs and stand-alone devices to decode CW. Some of the newer radios also decode. They work best with strong, clear signals and perfectly-sent computer-generated characters. If you're not proficient at CW, the decoders make the difference between a contact and no contact.

One good software decoder is called CW Skimmer by VE3NEA and available at dxatlas.com/cwskimmer. It works with your computer soundcard and displays a waterfall, [24] callsigns, signal reports, and text scrolling across the screen. Skimmer can decode multiple signals and is a very powerful tool.

---

[24] A waterfall display shows signals in two dimensions: time and strength.

MODES

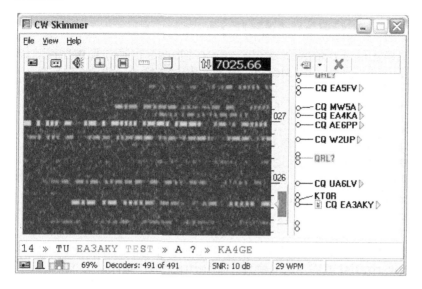

In this screenshot, the program displays about 3 kHz of spectrum. The waterfall is running sideways from right to left. Stronger signals are more intensely colored.

Callsigns extracted from the decoded CW are displayed on the right with a notation if they call CQ. The signal in the passband identified by the bar below "026" is decoded across the bottom. The program also listens for and displays signal reports, although there are none in this screenshot.

Skimmer comes in handy when the DX is working split. The calling stations will be spread out over several kHz. To tell where the DX is listening, tune the DX station into the decoder section at the bottom of the screen. Then, watch for callsigns as he works the stations calling him above his transmit frequency. Those who work the DX will be identified by the 599 signal reports displayed to the right. The latest 599 will be in red, showing who the DX just worked. It just so happens, the screenshot does not show any signal reports. They are usually there.

MODES

CW Skimmer does much more, but this is not the place to go into all the program's intricacies or how to use it. Software like this is complex and changes. Any details may be out-of-date by the time you read about them here. My job is to show The Easy Way. Your job is to implement it.

Compared to SSB, CW has a narrower bandwidth with a transmitted signal approximately 150 Hz wide. That means 20 CW signals can fit in the same space as one SSB signal. The typical filter used for CW is 500 Hz wide, although many receivers can go narrower. The narrower filter cuts down on interference and noise outside the passband.

Search using a wide filter so you can hear more and don't tune past a station. Switch to a narrow filter once you find a signal. Copying is all about the signal to noise ratio, and anything that reduces noise around the signal (outside the filter passband) will improve the copy.

CW has a significant advantage over SSB in that a CW signal is readable 10-13 dB below the level of an SSB signal. A good CW operator can copy at the noise level or slightly below, which is just about impossible with SSB. Since every 10 dB represents ten times the power, and another 3 dB doubles that, a 100 watt CW signal can be the equivalent of a 1,000 – 2,000 watt SSB signal.

I also believe many DX stations operate CW because they do not want to deal with the crowds on SSB. If you're not on CW, you will miss them.

MODES

DIGITAL MODES (PSK, FT8, RTTY)

There are many digital modes, but I limit this discussion to PSK, FT8, and RTTY because they are the most common modes for DX. Digital modes rely on a soundcard to generate and decode a warbling audio signal. Software converts the received signal to letters.

PSK caught on quickly after introduced in 1998. FT8 is newer and taking the world by storm. PSK and FT8 are very narrow-banded, and the decoding software is robust, so they can operate at very low signal levels. Both can decode a signal too weak to hear and barely seen on the waterfall display.

There are many free PSK programs available, and they do not require a super-computer to operate. Some work with a tablet. FT8 uses a free program called WSJT.[25]

The audio hookup from computer to radio is not complicated. For transmission, feed audio I from the computer to the microphone or line-in jack on the transceiver. The audio output from the transceiver goes to the computer soundcard input. Interfaces provide isolation to prevent ground loops and RF feedback. Check out the reviews for "Interfaces" on eHam.net or roll your own. Many newer radios have a built-in soundcard and use a USB connection to the computer to handle audio.

FT8 sends and receives in 15-second windows. It is not a "chatty" mode, limiting messages to a signal report, grid square, and confirmation. The software decodes multiple signals, identifies the stations, and responds with the appropriate exchanges automatically. The transmitter stays on one

---

[25] Weak Signal Joe Taylor, named after the author, K1JT

MODES

frequency, and you pick a "hole" in the waterfall display to transmit. The program sets your audio tone to fit in that hole. The following are a series of screenshots demonstrating FT8.

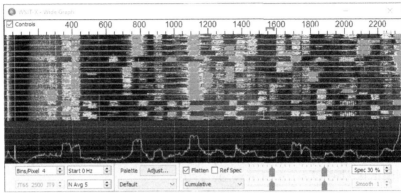

FT8 Waterfall and Spectrum Display

The marker above the waterfall, at around 1600 Hz, is where I will transmit.

FT8 Main Screen

The control panel shows stations calling on the left. CQ is in green. Red identifies someone calling me.

MODES

On the right are stations I am receiving. The band is busy.

JTAlert

JTAlert decodes multiple signals, identifies the ones calling CQ in green and LoTW in yellow. Jeffrey, AJ6IY identified on the left, is my QSO partner. The lower right is the status of my awards.

With narrow bandwidth, robust decoding, and the ability to decode weak signals, it is easy to see why digital modes are popular among DXers. The modes are too slow for running big pileups, but a new implementation of FT8 promises to allow multiple contacts simultaneously. A DX station could work everyone in the pileup at once.

RTTY, or teletype, is another soundcard mode that relies on computer power for encoding and decoding. RTTY is faster than PSK and FT8 and occupies a broader bandwidth

RTTY will operate with many free programs. You will almost always see RTTY signals on the bands, and there are several very active RTTY DX contests during the year. RTTY is popular with DXers, and should be in your arsenal. Don't be that guy who missed North Korea because he wasn't set up for RTTY. You'll meet him in a later chapter called "Fish Stories" – as in "the one who got away."

WHO'S ON?

Not that long ago, DX chasers arranged elaborate telephone trees to pass the word when a DX station came on the air. Today, there are better ways to get DX news.

MAGAZINES

QST and *CQ* magazines feature alerts to upcoming activities. Because of the lead time in publishing a monthly magazine, these alerts are usually speculative teasers. Still, they are a worthwhile "heads up" for the future.

ONLINE AND BULLETINS

ARRL publishes free weekly email newsletters for propagation and DX. Instructions for subscribing are on the ARRL website.

An Internet search will reveal many other sources of DX news. 425DXNews has a weekly news bulletin on its website. DX Publishing offers a weekly bulletin either mailed or emailed to subscribers. DX Summit has live spots and articles. DXZone has a vast collection of ham radio links, including links for DX information.

Another great source of information about upcoming DXpeditions, contests, and all things DX is at ng3k.com.

SPOTTING NETWORKS

Spotting networks allow stations to report what they hear. VHF packet carried the first spotting networks. Today, like everything else, spotting networks are on Al Gore's amazing Internet.

WHO'S ON?

There are stand-alone spotting resources such as DX Summit and others. Many logging programs integrate a spotting function. DxLab allows you to select multiple spot sources and displays the results in the SpotCollector window. Contest logging program N1MM+ collects spots and displays them in a band map window. Point and click sends the radio to the frequency. Spotting makes working DX like shooting fish in a barrel. Therefore, spotting networks qualify for inclusion on the list of Easy Way solutions.

Spotting is an invaluable tool, but use it with caution. Busted calls are common. Always verify what you hear versus what is spotted. As always, listen before transmitting. The spot may reveal the DX is working split or is only working Asia now. Calling out-of-turn or on the DX frequency will commit two mortal sins and get a well-deserved round of abuse from the frequency police.

Upload spots to the network, along with information about his split frequency and any remarks you might add. Go easy, and please don't whine about the operator or the unruly pileup. Everyone listening already knows, and adding to the negativity doesn't help. Your comments might make it worse as others feel justified to act out their frustration.

Never spot a station until you have worked him. Don't add to your competition. And, consider whether to spot him at all. If the pileup is already over-excited, it may be best to say nothing. Give the DX a chance to whittle down the mob.

If the DX is nothing special, don't clutter up the network with meaningless spots unless adding something new. Spotting the first EU station heard as the band opens is OK. Spotting a German station for the umpteenth time is not.

WHO'S ON?

A lonely CQer might appreciate a spot. DX stations might ask, "Please spot me." That is OK, but self-spotting is not. Imagine the chaos if everyone who called CQ spotted themselves. The network would have too many spots; it would be overwhelmed and worthless.

Spotting networks make it easy to find DX, but do not let the crutch become a wheelchair. Once a spot appears, the pack of hounds zeroes in quickly, and the advantage shifts to mega-stations who can overpower you in the pileup. You can still make contact, but it will be a lot harder. Ideally, find the DX before the rest of the world. That means lots of time carefully listening while tuning the bands.

Sovereign Military Order of Malta
I don't think they were /AM – Aeronautical Mobile.

DX FREQUENCIES

Where's the DX, you ask? It could be anywhere, but there are frequencies designated as DX windows, part of a voluntary band plan to leave open space for the exclusive use of DX. It is bad manners to conduct domestic rag chews or call CQ in these windows. The DX window is a place to listen.

The DX windows for Region 2 (that's North and South America) are:
1.830 MHz – 1.840 MHz CW
1.840 MHz – 1.850 MHz SSB
3.500 MHz – 3.510 MHz CW
3.775 MHz – 3.800 MHz SSB
7.000 MHz – 7.025 MHz CW
7.175 MHz - 7.200 MHz SSB
14.000 MHz – 14.025 MHz CW
21,000 MHz - 21.025 MHz CW
28.000 MHz – 28.025 MHz CW

Notice the CW windows on 40, 20, 15, and 10 meters are the entire Extra Class portion of the CW band. Everyone recognizes that's a bit too much, and the observed window is more like .020 - .025. Both phone and CW DX are usually right around the line between the band's Extra and General Class portions. Go there, and you will find them.

It is common for DXeditions to transmit right below the General Class band and listen up in the General Class band. As long as you transmit on the split frequency in the General Class band, your General license is good even though the DX was in the portion of the band reserved for Extras.

I think of those frequencies at the bottom of the General Class bands as the catbird seat. For instance, 7.026 – 7.030 MHz is a great place to listen for or call CQ DX. Always listen first and ask if the frequency is

DX FREQUENCIES

in use (QRL?) before calling CQ. Be prepared to be shooed off if other DX activity begins.

No one owns a frequency, but don't be a frequency hog. It is easier to move and maintain peace than to generate animosity and strife. When asked to move, move at the first convenient time. Consult the ARRL Considerate Operator's Guide available on the Internet.

Here's a peaceful looking spot to call CQ.

TRANSCEIVER SELECTION

I am not going to recommend particular models or manufacturers. We all have our favorites, and some people like Fords while others favor Chevys. As my beloved Nana used to say, "That's why they don't make all ice-cream vanilla." The transceiver has taken the place of discrete transmitters and receivers. I will discuss the criteria to apply when choosing a modern transceiver for DX.

TRANSMITTER

The transmitter section is not nearly as critical as the receiver. The receiver will dictate your choice.

Fortunately, this is not a dilemma. Most transceivers deliver 100 watts, and 100 watts is the same no matter how it comes. Okay, one can argue about phase noise, clean signals, and key clicks, but modern equipment has pretty much slain those dragons unless you operate near other Hams.

Some transceivers output 200 watts. Are 200 watts necessary? Doubling power only increases the signal by 3dB, or one-half an S unit. Decide if the extra money is worth it, but I doubt if anyone will be able to hear a half S-unit difference. A beefier amplifier or antenna is a better use of your money.

Look for a built-in speech processor to tailor the audio, as we discussed in the Mode chapter. A voice and CW keyer would be nice, but computer software or outboard devices can implement those functions.

TRANSCEIVER SELECTION

RECEIVER

The receiver is the most critical part of a DXer's station. Sensitivity is not the issue as they are all about the same and will receive at the noise level. If you can hear atmospheric noise, the receiver is sensitive enough. Consider reducing the RF gain.

What sets an excellent receiver apart from a mediocre one is selectivity and, most importantly, selectivity to resist intermodulation by-products (IMD) from strong signals mixing.

The latest tool to defeat such interference is called a "roofing filter" applied very early in the receiver chain to reduce the breadth of signals reaching the later, more sensitive circuits. Some radios have selectable roofing filters as narrow as 250 Hz. A narrow roofing filter will also reduce AGC (automatic gain control) pumping caused by loud signals in the close-by pileup.

I once had a relatively high-end radio with a 15 kHz wide front end. On a crowded band, I could hear what sounded like a growling noise in the background. It didn't matter that I switched to a narrow crystal filter; the distortion came from mixing signals – Intermodulation Distortion (IMD). It was fatiguing, distracting, and made weak stations almost impossible to copy. I replaced that radio with another having a much narrower front end, and never heard the growling again. A narrow front end is best.

In addition to a roofing filter, there is filtering down the chain using crystal filters or Digital Signal Processing (DSP). The receiver should have filters appropriate for the mode. Some transceivers ship with one moderately wide filter for SSB, say 2.7 kHz. Add a narrow 1.8 kHz SSB filter along with a 500 Hz CW filter. DSP filtering is usually continuously adjustable, not limited to preset crystal filter bandwidths.

TRANSCEIVER SELECTION

A transceiver should implement some form of split mode and preferably one that allows listening to two frequencies simultaneously. Many transceivers have a VFOA and a VFOB but implement them differently.

An ideal set-up is a transceiver with two independent receivers, so the DX is in one ear of the headphones and the pileup in the other ear. Tune around on the second VFO, listening in that ear for the station who just worked the DX. That tells where the DX is listening. With the transceiver set for split operation, your call will go out on the second VFO frequency, not the DX frequency.

An alternative to two receivers is sometimes called "dual-watch." Dual watch plays to two frequencies simultaneously, but they are mono, and there is no physical separation of the fox and hounds. Another alternative is to press a button and listen to the second frequency. The button-pressing rhythm is critical not to miss the call. Two receivers are the best implementation.

Another useful feature is an easy way to access the IF frequency output to feed a panadapter. Many transceivers now have an IF output jack or a panadapter display on the front panel. That is preferable to hacking into the radio to get the IF out yourself. The next chapter discusses panadapters.

PANADAPTERS

A panadapter shows a range of signals on a visual display. Many modern transceivers feature a panadapter on the face sharing the control panel screen or available through a video output to a separate monitor.

Stand-alone panadapters feed the receiver's Intermediate Frequency (IF) to a Software Defined Radio (SDR) receiver and use software to display the results on a computer monitor. The SDR can be as straightforward and inexpensive as a USB dongle if you want to "roll your own."

One excellent implementation of this technology is the LP-PAN by N8LP, Larry. The LP-PAN device is reasonably priced and produces a beautiful display of up to 192 kHz bandwidth when used in conjunction with a quality soundcard. The soundcard in your computer may not support that much bandwidth, but outboard sound cards have come down to a very reasonable price point. Larry has suggestions on his website, TelepostInc.com.[26]

A program called LP-BRIDGE allows the radio, logging program, and panadapter to communicate over a single USB cable by setting up "virtual" serial ports to control each function. The SDR software and LP-BRIDGE are free.

I am describing functionality, not endorsing a particular product. There are other panadapter solutions, and my mention of LP-PAN is not intended to disparage any of them. Elecraft has the P3, and newer radios have panadapters on the front panel.

---

[26] I have no affiliation with LP-PAN or any other product mentioned in this book.

PANADAPTERS

This screenshot from an LP-PAN and NaP3 software demonstrates why using a panadapter is another example of The Easy Way.

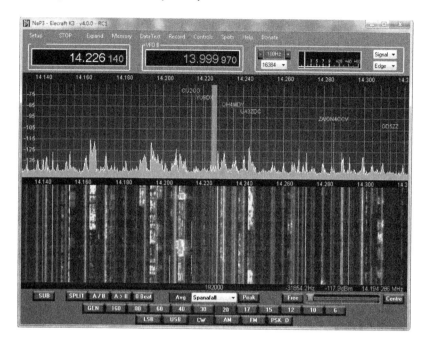

The upper part of the display is the spectrum and shows almost the entire mode-band. The lower display is a waterfall showing back in time. Click on a signal, and the radio shifts to the frequency.

One glance at the panadapter tells if a band is active. Signals are visible, or not—no need to tune across the entire spectrum to find out. The display can also identify a pileup and show where the open frequencies might lie.

In the screenshot, spots coming in from an Internet source superimposed on the display show not just the signal but also who it is (subject to the usual caveat of double-checking spots for busted calls).

AMPLIFIERS

Some folks work their DX using low power, QRP. My hat is off to them. I have done it myself, including this two-way QRPP (very low power) QSO with LZ2RS in Bulgaria. Rumi and I were both running 100 milliwatts (1/10 of a watt) for 49,500 miles per watt. Notice my signal report of 509. That may be a fun way, but it is undoubtedly not The Easy Way.

How much power is enough? Doubling power adds 3 dB or one-half S unit. Going from 100 to 200 to 400 watts adds one S Unit. 400 to 800 is another half, and 800 to 1600 only adds another half. Can anyone hear the difference between S6½ and S7? That is why it is often said, "The first 400 – 800 watts are the most important."

Running full power stresses everything in the station, from the equipment to the antenna tuner to the coax to the antenna. The more power run, the more likely you'll generate RFI in the shack and have the neighbors complaining about their garage door going up and down.

AMPLIFIERS

Big power requires proper planning and a large budget. Every dB gained makes the DX job easier. But, getting back to the original question, "How much is enough?" The obvious answer is, "As much as you and your wallet can handle."

Science says the first 400 – 800 watts are the most important, and I will stick with that answer. Many tube and solid-state amplifiers in this range will serve well and not break the bank or station. These mid-range amplifiers should give plenty of fire to get DX in the log. Five-hundred watts will run more juice than most of the competition and be less than an S unit below a 1,500-watt big gun.

Depending on the rest of the circuit load, a 500-watt amp may operate from a 110V power outlet. That saves the extra work to get 220V to the shack. I was fortunate that the electric dryer outlet was accessible nearby, and I could tap into it to run 220V to my amplifier. It is best to run 220V. The amplifier will appreciate it. Ohm's Law says the current draw will be one-half, and the voltage drop one-quarter, making component-stressing and distortion-inducing voltage sag less likely.

The choice of tube vs. solid-state usually boils down to price. The pros for solid-state amplifiers are instant on, no twiddling of knobs to tune, and instant band changes. The cons are they do not tolerate SWRs over 2.5:1, so they might need an outboard antenna tuner. We're twiddling knobs unless you get one of the newer auto-tuners. They are quite good, and the cost is not much more than a manual tuner, but it is an additional expense added to an already expensive amplifier.

Tube amplifiers are much more tolerant of higher SWR because the plate and load knobs (tank circuit) adjust for tuning. The tank circuit is a form of an antenna

AMPLIFIERS

tuner. Solid-state is considerably more expensive than tube, and tubes tolerate abuse better. A blown transistor is harder to replace than a blown tube.

I've had both, and there is no "right" answer for this one. On a budget, I would go with a moderate tube amplifier and put the money saved into a better antenna. A better antenna will make a difference on both transmit and receive. You don't want to be an alligator – "all mouth and no ears."

This Spanish station has some serious ears.

ANTENNAS

There is probably no more contentious area of discussion than the choice of antennas. Remember, "If you ask five hams a question, you will get seven different opinions and probably a fistfight."

A better antenna helps on both transmit and receive, so the antenna system is critical. Not everyone can erect a monster tower equipped with stacks of monobanders. Many Hams are antenna challenged. Do the best you can and get on the air.

A little-pistol may not get the DX on the first call, but finesse is far more valuable than brute force. Being an S unit or two below the strongest station doesn't matter. I've heard many pileups where weaker signals get the prize. Learn the proper techniques to will get the contact.

Everyone agrees that gain is good. So what about gain? All antennas have gain relative to something, and advertisers brag about their antenna's gain. The question is, "Is the gain in the right direction?"

VERTICALS

Some say, "Verticals radiate equally poorly in all directions." I worked tons of DX with a vertical on my roof. My yard was too small for anything more, and I didn't know any better. Verticals work best on low-angle radiation – like the signals that come in from far-off DX. That is where they have "gain." A properly-erected vertical can be a good DX antenna. That means lots of radials if ground-mounted and a few resonant radials if mounted above ground. Some verticals require no radials at all. They are designed as vertical dipoles or have matching networks to simulate radials.

ANTENNAS

HORIZONTAL DIPOLES

Horizontal wire dipole antennas radiate off their sides, and that is where they have gain. If operated on frequencies above their fundamental, they develop lobes – first a four-leaf clover and eventually a multi-petal flower shape. Use a modeling program like EZNEC[27] to predict performance.

Multi-band capability is important. Be where the DX is. Multi-band designs like off-center-fed Windoms and ladder-line-fed doublets can work DX. Other multi-banders include fan and trapped dipoles.

The pattern for horizontal antennas hung less than a half-wavelength high tends to become omnidirectional with a higher angle of radiation, so get them as high as you can. Maybe erect a vertical and a horizontal antenna. There is no one-size-fits-all solution.

ROTATABLE ANTENNAS

Of course, you know where this is heading (pun intended). Horizontal antennas are directional, and it would be useful to turn and radiate in the desired direction. Signal reduction off the end of a dipole can be 30dB (5 S-units).

A rotatable dipole (one element beam, if you will) is a significant improvement over a fixed dipole. If the DX is off the end of an antenna, redirect the antenna. A rotatable dipole is not too expensive and is light enough to mount on a roof and turn with a TV rotor. It will reject signals off the ends but has no front-to-back rejection and only a modest gain of 3 dB or half an S unit over a vertical.

---

[27] A free demo version is available at EZNEC.com

ANTENNAS

Step up to a two-element Yagi and increase the gain plus reject signals off the back. Front-to-back and front-to-side rejection reduce interference and noise from other directions. A two-element Yagi is a significant improvement over a dipole and is an excellent trade-off between gain and expense. An additional 4dB gain more than doubles the effective radiated power.

Since antenna performance is reciprocal, the same when receiving as transmitting, gain will also increase received signal strength. The incremental benefit of a 2-element Yagi over a dipole is substantial and probably the best value found.

Adding a third element increases the gain and front-to-back rejection but with diminishing returns. The additional antenna element only adds 1-2 dB gain and maybe 5 dB of front-to-back. The antenna is bigger, heavier, has more wind-load, and will require a more robust rotor and support. The 3-element Yagi is very popular, and you won't regret getting one. But, on a monetary or engineering budget, the 2-element version is a good compromise.

To know which way to turn that directional antenna, use one of the azimuthal map programs on the Internet. One source is QSL.net/VE6YP. Print out a map and keep it nearby. DxLab's DXView module displays bearings from your location. See the Logging Programs chapter.

Work with what you can and hone your skills. Finesse is more important than power. Any antenna can work a lot of DX. The perfect antenna you are looking for but still haven't installed won't work anyone!

SHACK DESIGN

Having seen many truly horrible shack setups, I feel compelled to add my two cents about design and layout. If you are going to spend time chasing DX, make the operating position as comfortable and inviting as possible.

I shudder to see hams relegated to windowless, damp, and dingy basements crouched among the cobwebs, behind the furnace, out of contact with the rest of the household. Your shack is your pride and joy, so come out of the cellar. Spouses don't hate your hobby. They hate your isolation.

You can't make a lot of noise in the living area, but isn't there a room somewhere on the main level to get away from the TV and have a little solitude? How about a den or sun porch? I like to look out the window and see the birdfeeder or the grandkids playing.

No matter where, brighten it up with adequate lighting and a rug on the floor. Get a comfortable desk with room to spread out. I used a door on sawhorses as a kid. Now, office furniture is available in comparatively cheap, attractive, and utile packages. Get a decent office chair and stop sitting on that old broken, and unpadded kitchen stool. Make yourself comfortable so you can be a part of the family and maximize your BIC time without killing your attitude, your posture, or your marriage.

The computer is a significant element in the modern ham shack. The monitor(s) and keyboard should be front and center. An off-center keyboard or monitor requires twisting your body leading to neck, shoulder, and back problems. Two 23-inch monitors display my logging program, spotting reports, panadapter, rotor control, etc. Stacking them one above the other was

SHACK DESIGN

straining my neck. Now, the monitors are side-by-side, and I can see both with minimal body movement

Sit up straight. Do not hunch over the keyboard, raising your shoulders above where they hang naturally. To avoid hunching, lower your chair, or raise the keyboard.

Adjust the monitor height, so eye level is slightly above the center when sitting up straight. We see better "down" than "up." Look down a hallway and notice you see more floor than ceiling.

Try shifting the mouse to your "off" side occasionally. Reverse the buttons in the computer's mouse control application. The change is good for your body alignment. Pain in the shoulders, neck, or back is evidence of doing it wrong. Trust me, I know. A couple of years ago, I had a severe case of "frozen shoulder." I could not raise my arm, and it took many months with a chiropractor to work out the kinks.

Often-adjusted equipment should be within easy reach. The transceiver should be centered or slightly off-center to favor your preferred hand. The radio belongs at the desk level to rest your arm while operating. Reaching for a radio above the desk is very tiring and uncomfortable. My Morse key is to the right as I am right-handed.

I have two small speakers on either side of the radio. They provide excellent stereo sound. For split operation, DX is in the left ear, the callers to the right. Stereo is good for more than split operation. Many transceivers allow for phase reversal or other audio effects that enhance the sound giving the impression the signal is in your head, not your ears.

Don't waste money on speakers that match the transceiver. "Matching" only means they paint them

SHACK DESIGN

the same color as the transceiver. I am pleased with Pyle Home mini-cubes available for less than $25 a pair. They are about four inches square.

Close-up of the K4IA operating position. The microphone is mostly for show.

When the going gets tough, I pull out the cans (headphones). Don't spend a fortune on high-fidelity studio-quality equipment because deep bass and tinkling highs don't matter. Everything you want to hear is between 400 and 3,500 Hz. The most critical specification for headphones is comfort.

Over-the-ear designs cover the ear entirely like ear-muffs and block out external noise. They can be uncomfortable, sweaty, and clammy for extended wear.

On-the-ear designs rest on the ear and don't provide much in the way of noise-blocking but are more comfortable. Earbuds fit in the ear and do a better job of noise blocking than over-the-ear designs. I have all three and change them around to suit the conditions or comfort.

SHACK DESIGN

Some headphones are "noise-canceling." They sample the external noise in your shack, reverse its phase and feed it back into the audio system. Your ear perceives the noise as gone. Noise-canceling headphones reduce shack background noise, not noise on the signal.

The desktop can get crowded. A boom microphone won't take up desk space, and a footswitch frees your hands. I recommend foot pedals designed for tattoo parlors. They are rugged, inexpensive, and available online. You will have to attach the proper plug for your radio.

Shelves can hold auxiliary equipment such as the antenna tuner, wattmeter, and digital interface, all within sight and reasonable but not immediate reach. I don't adjust those components often, but I want to see the displays.

As a general rule, anything that does not need adjusting can be out of reach. The rotor control box is in sight but out of reach. The rotor control application in my logging software turns the rotor, so I don't need to spin the dial manually.

The power supply is on the floor where it belongs. I use my big toe to turn it off and on. Putting the power supply next to the transceiver can induce hum and other noise. It also takes up valuable space. You never adjust the power supply, so why should it clutter the desk?

The computer tower also does not need to be on the operating desk. When was the last time you adjusted anything on the computer?

COMPUTERS

Radios and computers go together, like peas and carrots. The modern transceiver interfaces with computer logging programs, digital modes, CW keyers, voice keyers, rotor controls, panadapters, Internet spotting, callsign lookups, and QSLing. A computer and radio interface is essential.

There is some software written for Macs, but the majority is PC oriented. Very little runs on Linux. Chromebooks rely on access to the Internet to run programs, and I am not aware of any ham software that will run on a Chromebook. Experimenters are always pushing the envelope, but the vast majority of ham software is written exclusively for the PC.

An expensive high-powered computer is not required. Most ham software will run Windows 7 or above. Support for XP is gone. Adding memory is the most cost-effective way to improve computer performance. Visit Crucial.com and run their scanner program to find the memory upgrade options for your machine.

The computer can be out of the way, but bond the case to the shack ground and other equipment with a grounding strap.

"Bonding" provides a parallel low-impedance path for RF to flow from chassis to chassis and ground. Toroid chokes on the interconnecting cables raise their impedance, keeping RF out of the internals. The combination of choke and the low impedance parallel path between components' chassis avoids RFI (Radio Frequency Interference). Here is an excellent article by K9YC, Jim, and it should be required reading: audiosystemsgroup.com/RFI-Ham.pdf

A wired or wireless Internet connection provides the maximum benefit of downloading spots, looking up

COMPUTERS

callsigns, backing up the log, and uploading logs to LoTW.

The newest radios have a direct USB interface to the computer. However, many older radios use a serial port interface, but newer computers lack serial ports. Install a serial port card in a desktop computer if it has slots for such a thing. Serial port cards are cheap.

A USB to serial converter cable/box can feed a radio's serial port. Those are inexpensive as well. ProLific and FTDI converters had issues with counterfeit chips, and both companies inserted code in their drivers to disable unlicensed fakes. Best not to buy cheap no-name knock-offs on the Internet.

The computer can also act as a voice or CW keyer. For voice, sound from the computer feeds into the transceiver. The newest radios have a built-in sound card. If yours doesn't, there are many devices designed to provide isolation and trigger your PTT. Check the "Interfaces" reviews on eHam.net.

On CW, the old way used a serial port to trigger CW. Serial ports have timing issues, so the modern solution is a USB device. Visit K1EL.tripod.com for the WKUSB, an inexpensive CW keyer, or the eHam reviews for "Interfaces" and "Keyers and CW keyboards."

Another chapter covers logging programs, but the combination of a computer and a logging program is essential to a successful DXer.

WHERE CAN YOU OPERATE?

If you operate away from your home station or use a remote-controlled station, contacts can still count toward DXCC credits as long as the transmitter is within the same DXCC entity. A move from DC to LA doesn't start the count over, and when visiting on the West Coast, stations worked from there toward the DXCC totals.

Advances in Internet technology have made remote control stations possible. Not long ago, I spoke with an amateur using a callsign in the Azores. I recognized his voice, so I asked: "Hey Martti, how's the weather in the Azores?" Martti came back, "I don't know. It is cold and snowy here. I am sitting in my living room back home in Finland." Martti was operating his remote station in the Azores over the Internet. That's pretty amazing, considering the operator was thousands of miles away from the equipment. The control point was in Finland. My friend in Finland operating his station over the Internet was an example of remote control. He was twiddling the knobs from afar.

WHERE CAN YOU OPERATE?

Here's the quote off the back of Martti's card:
The technology is here to overcome noise and restrictive antenna ordinances and help you set up a shared station at a reasonable cost with your friends. Or, you can even operate from distant lands without being there.

You have now contacted CU2KG Remote in the Azores, with the station operated through the Internet from Finland. You have contacted a highly advanced station employing the latest technology and its people. Just come on board for the thrill of being another remote station, using the Internet as your link.

Many modern radios offer remote control access and software to operate the home station while away on vacation or from the office. There is at least one group offering remote control access to super-stations featuring amplifiers and massive antenna systems. They charge a subscription fee and a per-minute fee. Remote operation sounds like an excellent solution for Hams constrained by Homeowner Association rules or lacking their own equipment. Not everyone has a super-station or gets to erect a 90-foot tower.

"Do remote operations count and from where? Do we consider the operator's or the transmitter's or the antenna's physical location?

In 2015, the ARRL adopted new rules concerning remote control operations. The rule says the transmitter's and antenna must be co-located (together), and the transmitter's location determines the country of origin. If the transmitter is in the Azores, it doesn't matter where the control operator is sitting; the contact counts as one from the Azores.

If an American Ham is trying to add to his USA DXCC count, a contact made from a transmitter in the Azores won't count since the transmitter is not in his

WHERE CAN YOU OPERATE?

DXCC entity, the USA. However, he could control a remote station somewhere else in the USA, and those contacts would count for award credit.

Now, the question becomes, "If I operate a remote station out of my country, using my US callsign, how does anyone know?" Using a foreign transmitter while identifying with a US callsign is breaking the licensing laws. The laws and rules require a license or permission in the country of transmission and that you correctly identify as CU/K4IA, for example. But what if you don't? Who is the wiser?

Here we come to the question of personal integrity and responsibility. I do not for one second condone breaking, stretching, or bending the rules as "The Easy Way." Awards earned by cheating are worth nothing, and you will always wonder if it could have been different.

Don't compromise. Play by the rules and obey the law. That DXCC certificate or Honor Roll plaque on your wall should mean something. You can't be proud of an achievement after cheating to get it.

Remote operation is an evolving field. Watch for new developments in equipment and operating rules.

CONFIRMING YOUR CONTACTS

A contact requires confirmation for it to count toward most awards. The QSL card is one way to confirm a contact. Logbook of The World is another. *CQ* magazine recognizes eQSL for some awards. LoTW and eQSL are online services. The major awards do not recognize other online QSL systems at this time.

DXCC (DX Century Club) requires confirmation from 100 entities. The 5BDXCC (Five Band DXCC) will take 500 confirmations, 100 on each of five bands. 8BDXCC needs 800. Assuming a 50% confirmation rate requires sending out twice the number of cards you receive. That is a big job. Pity the poor DX operator who must deal with hundreds of cards a month. He may spend more time answering cards than he does on the radio, and the expense of printing and postage is considerable. If the DX operator is picky about how he answers QSLs or wants a few Green Stamps, understand and comply. He is under no obligation to humor your request.

QSL CARDS

The old saying goes, "A QSL card is the final courtesy of a QSO." QSL cards have been around since the beginning of ham radio. I love collecting them, and you've seen some of my favorites on the pages of this book. I have shoeboxes full. It is fun to reminisce while looking at cards from exotic lands with beautiful pictures, strange handwriting, and unique designs.

You can go cheap with yours, but printing QSL cards is not a significant expense. Pick something interesting enough to elicit a response. A plain-Jane card is just as good as a fancy one for confirmation purposes, but I like to send something nice, so the other guy feels like I cared enough to deserve his card in return.

CONFIRMING YOUR CONTACTS

Cards with your picture or a picture of a local landmark or a cartoon convey a personal touch. Unique cards are interesting to receive, and I believe they are more likely to generate a response. Who can resist answering this card from Northern Ireland?

A custom card can commemorate a memorable contact – like this one from Australia. I appreciate the extra work Chris put into making a unique card for me.

CONFIRMING YOUR CONTACTS

Here is an example of a very simple home-made card from Russia. It is a rubber stamp on cardstock. I like this card because it is so retro.

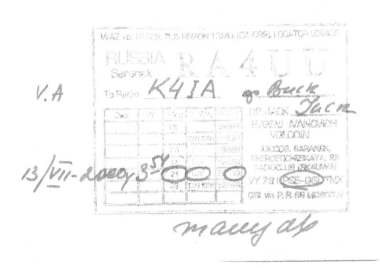

As nice as they are, I didn't find self-printing cards cost-effective or worth my time. I was giving up valuable radio opportunities fussing over artwork and printer issues. There are many professional printers offering full-color and picture QSL cards. They will do the job cheaper and better than your buying cardstock, ink, etc.

Fill out your card legibly. It doesn't help if the reader can't tell the date or details because of your handwriting or ink smears. Logging programs offer the option of printing report labels to stick on the card.

Dates can be confusing. Is it month/day/year, or is it day/month/year like they use in Europe? What date is 5/7/21? May 7 or July 5? Spell the month in English. Write out the four-digit year: May 7, 2021.

If you write 5/7/21, and the DX looks in his log for July 5, you may not get a reply. Anything to make the

CONFIRMING YOUR CONTACTS

responder's job easier will pay off with a higher rate of QSLs confirmed.

Use a computer logging program, and you can be reasonably sure the contact information is transcribed correctly. If you are hand logging, check and double-check the data.

Here is the card I am using at the moment. It comes from QDure, and I'll have more to say about them later.

QSL DIRECT

Once you have your card, you need to get it to the DX station and receive a confirmation in return. Now the fun begins.

You can send cards directly to the DX after getting address information from QRZ.com or other callbook services. Follow the DX's instructions. Some DX stations specify they will only QSL direct. On CW, they

CONFIRMING YOUR CONTACTS

might say, "QSL CBA," meaning "QSL to my Call Book Address."

If you QSL direct, assume the DX expects a few "Green Stamps" for the return postage. Green Stamps are dollar bills. Understand the cost of printing and mailing QSL cards can be staggering for a foreign ham. Overseas postage is $2 - $3 from Europe. Most DX operators are not "dialing for dollars." "I am DX. Hear me holler. Send me dollar. Send me dollar."

Don't send American postage stamps because they aren't suitable for postage from a foreign country. The only time to send American stamps is if the DX is a stamp collector and asks for them on his QRZ.com page or elsewhere.

It takes at least $2 to cover the cost of return postage from another country.[28] Add $1.20 for your postage, and QSLing direct gets expensive. I would reserve it for exceptional cases.

Other than the expense, the problem with sending cash through the mail is that it will be stolen. Postal thievery is a big problem in some parts of the world. The bad guys recognize envelopes that may contain cash and steal them. Even a few dollars a day adds up to real money for them.

Disguise the envelope when sending cash. Do not put his callsign, your callsign, or anything related to Amateur Radio on the envelope. Make the envelope appear to come from a business. I use a fake company name in my return address. Something like ACME COLLECTIONS or THE BUCK LAW FIRM doesn't

---

[28] An International Reply Coupon (IRC), which you could buy at a US Post Office, was good for return postage anywhere in the world. They are being phased out and are not accepted universally. Don't waste your time and money on IRCs.

CONFIRMING YOUR CONTACTS

scream "money inside." You might also print "Invoice Enclosed" on the envelope, so it looks like a bill and not a dollar bill is enclosed. That would be "Factura Incluida" in Spanish or "Facture Jointe" in French.

Use a security envelope – the kind with shading on the inside. Don't make your envelope lumpy. Use a printer or label, and don't hand-write the address.

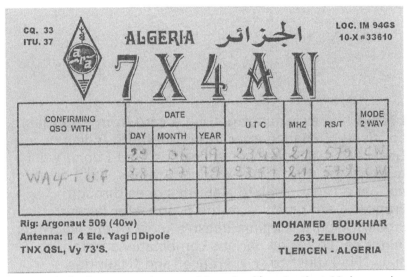

Postal theft was such a problem in Algeria that Mohamed asked for registered mail. Thankfully, he is now QSLing through a PO Box in Barcelona, Spain.

Use a self-sealing envelope for the return. Take pity on the unfortunate DX who has to lick a hundred envelopes.

Make sure you add "USA" to your address on the return envelope. Georgia is a country, as well as a state. A foreign postal worker may not recognize your address as being from the US. Your reply will end up in a dead letter bin, along with all the letters to Santa Claus.

CONFIRMING YOUR CONTACTS

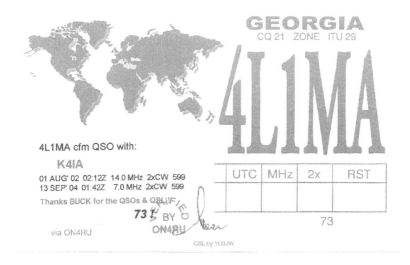

I had a bad experience getting a card from Mauritius, an island off Madagascar in the southern Indian Ocean. I sent dozens of envelopes, and I don't want to think about how many Green Stamps, but I could not get a reply. Mail to Mauritius goes through France. I am pretty sure the postmaster here in Fredericksburg, Virginia doesn't know where Mauritius might be, so I tried various combinations from just plain "Mauritius" to "c/o France" and "Mauritius, France." Nothing was working. I'm not sure if it was postal theft or the long route, but I could not get a reply.

I got lucky. I have a client from Mauritius, and his mother came to visit. She offered to carry my mail back to the island, put a local stamp on the envelope, add her Mauritian return address, and mail it from within the local postal system (not through France). I finally got my reply. We had so many QSOs that Jacky sent two cards covered front and back. I don't imagine something like that will ever happen again, but it makes a good story.

CONFIRMING YOUR CONTACTS

OQRS

OQRS stands for "Online QSL Request System." Bigtime DXers, including DXpeditions, really don't want your card – they have thousands if not tens of thousands. OQRS allows you to initiate a request online without sending out a card yourself.

ClubLog administers OQRS. A list of supported DXpeditions is at clublog.org/expeditions.php. The online logs confirm your contact, and you can request QSLs from the web page. Pay with PayPal for a direct card or request a QSL via the bureau free. There is often an option to donate toward the DXpedition's expenses.

When I first saw this, I thought it was sketchy – selling QSLs didn't seem to fit in with ham spirit or commercial hamming prohibition. I was wrong, and I've straightened out my thinking on this.

OQRS provides a way to guarantee your Green Stamps aren't stolen and saves you the effort of sending out a card. Returns via the bureau are

CONFIRMING YOUR CONTACTS

usually free, and most DXpeditions will upload to LoTW[29] six months to a year later. So, if you are willing to wait, you will get a confirmation eventually. The OQRS system provides a priority for those who want to receive a paper card directly and are willing to pay a small fee for the privilege. I am happy with that. I like a paper QSL for the rare ones.

My DxLab logging software automatically uploads QSO data to ClubLog. Registering with ClubLog is free. As of early 2021, there are over 717 million QSOs in the system and almost 83,000 callsigns participating.

QSL VIA A MANAGER

Some DX uses a QSL manager. The DX sends his log to a kind-hearted volunteer who manages the QSL duties for him. If you're fortunate enough to be dealing with a manager in the US, you send your card and a Self-Addressed Stamped Envelope (self-sealing) to the US QSL manager. With time, you'll receive your confirmation.

QSL managers are volunteers and may even pay for the card printing. Help them out by tossing in a dollar or some extra postage stamps. You will get on the manager's good-guy list, and that could help one day when you need a favor. It is the right thing to do.

Some managers handle multiple DX stations, so write the QSO information on the back of both envelopes (only for US managers). That will help sort incoming cards, and if your card and SASE get separated, the manager will be able to put them back together.

If the QSL manager is not in the US, you can QSL to the manager directly, but many will accept and return a QSL via their bureau. The route is to the manager's

---

[29] Logbook Of The World

CONFIRMING YOUR CONTACTS

QSL bureau, not the DX's. You put "His Call via Manager's Call" on your card. No SASE is needed because your return will come through the bureau. A few pages back, Georgia's card was via the bureau system and a QSL manager in Belgium.

Follow the operator's instructions or check with QRZ.com or another callbook service. If the DX is operating with his home callsign / foreign callsign (K4IA/4X4), you might hear him say "QSL to home call" or 'QSL HC" on CW. Send your card direct to his home address. "QSL via CBA" means QSL to the address specified in the callbooks – either direct or to a manager.

These Polish operators on Easter Island used a QSL route through the home call in Poland.

CONFIRMING YOUR CONTACTS

QSL VIA BURO

Another alternative is via the QSL Bureau system. Bureau is often shortened to "Buro" on CW. Here's how that works:

ARRL members can bundle up cards, sorted by country, and send them to the ARRL Outgoing QSL Bureau. ARRL charges a fee based on the weight or number of cards to collect the cards and forward them, in bulk with others, to the foreign country's QSL bureau service. From there, the foreign bureau distributes cards to its members. Not all countries participate, and that information and additional details are on the Outgoing QSL Bureau page of the ARRL website.

ARRL membership is not required to receive cards. For incoming cards, find the person who manages your callsign on the ARRL incoming QSL bureau website. Send him either stamped envelopes or money so he can purchase envelopes and postage. Contact the bureau manager to see what he prefers.

The bureau manager collects cards directed to you, bundles them together, and sends them when an envelope is full. It is like Christmas when that package arrives full of cards from all over the world.

This system is economical, but the round trip can be slow. Sometimes the replies come quickly, but I have had cards take years to reach me. I think my record is eleven years. Lots of DX uses the bureau system, and if you are active, you can count on a steady stream of beautiful and exotic QSL cards.

CONFIRMING YOUR CONTACTS

Speaking of Christmas, how about a card from Franz Josef Land, on the Arctic Circle, where Santa lives?

I think I would rather spend Christmas here, with Bert, on Christmas Island.

CONFIRMING YOUR CONTACTS

QSL SERVICES

A QSL service combines the best of the bureau system with computerized logging. Two companies offer the service: GlobalQSL.com, from Israel, and QDure (QSL.URE.ES/EN), sponsored by the Spanish equivalent of ARRL. Rumor is GlobalQSL closed in 2020, so I now use QDure. GlobalQSL may come back.

Here's how it works:

First, you go on the website and upload or design a card, front and back. They have templates and artwork available and will accept your pictures.

The next step is to upload an ADIF[30] file containing your outgoing QSL information. QDure prints that information on the card. No more labels or handwriting QSL data. No more sorting and mailing to ARRL. They warn if you try to send a card to a country with no bureau. I can upload hundreds of QSOs in a few minutes.

Cards are cued up and printed on demand. When enough cards are going to a particular bureau, they print them all and include yours with the bundle. No-fuss, no muss, and because they handle so many users, it doesn't take long for that package to go out.

QDure is doing all this for about thirteen cents a card. That price includes printing, personalizing, data entry, sorting, and sending to the foreign bureau. I uploaded 120 QSOs for $16 and had confirmation the order was complete in ten days. You cannot beat that.

---

[30] Amateur Data Interchange Format is a standard format used to insure interoperability among programs. Computerized logging programs produce an ADIF file with your QSL information.

CONFIRMING YOUR CONTACTS

You still need to participate in the ARRL incoming bureau to receive replies. And your return card can be slow to arrive.

Send the card design on QDure to your favorite QSL card printer and have them print your custom-designed card for stations that do not qualify for bureau services such as domestic contacts.

LOGBOOK OF THE WORLD (LoTW)

Paper QSLs can be expensive, even if you use the slow bureau system. A DXpedition or contest station might make tens of thousands of contacts. The burden of sending paper QSL cards is just too onerous. I love and prefer paper QSL cards, but I understand the problem.

An electronic confirmation cannot substitute for a real paper card from somewhere like Iran.

ARRL developed an online confirmation service called Logbook of The World. The premise is for stations to upload their logs, and the system matches contacts to create an electronic confirmation. Confirmations are

CONFIRMING YOUR CONTACTS

used to purchase award credits. The credits apply to ARRL awards and some *CQ Magazine* awards.

LoTW is free, but award credits cost 12 cents per credit. DXCC, requiring 100 credits, would cost $12 plus another $12 for the certificate to hang on your wall.

There is a bizarre double-encrypted registration process to maintain the integrity of the system. I won't go into the details, but it is all on the LoTW portion of the ARRL website. Have your Little Orphan Annie decoder ring handy. All kidding aside, ARRL has gone overboard to protect against data tampering. Read and follow the instructions carefully.

Once successfully registered and authenticated, upload logs from a logging program or by an ADIF file, and LoTW does the work. Most computer logging programs will upload automatically. There is no paper card, but confirmations come relatively quickly and for little or no cost. Compared to other methods, LoTW is lightning fast. Sometimes it takes a day or two for the system to process uploads.

LoTW is available for domestic as well as foreign contacts, but not everybody participates. I upload everything. When someone joins the system later, LoTW will match our contacts automatically.

As of early 2021, over 1 billion QSO records are in the system, and over 260 million QSL confirmations. I have 26,000 QSOs uploaded and about 11,000 confirmations. That is not a significant percentage, but it includes many domestic contacts for which I would not expect a QSL card. The money saved on those 11,000 confirmations is substantial.

Here is a LoTW report of my statistics. If I click on an entry in the far-left column, I get a report of my

CONFIRMING YOUR CONTACTS

confirmed contacts on that band or mode. Reports generate in a few seconds. I can't imagine the amount of time it would take to put this together manually. Logbook Of The World qualifies for special recognition as DX – the Easy Way.

DXCC Award	New LoTW QSLs	LoTW QSLs in Process	DXCC Credits Awarded	Total (All)	Total (Current)
Mixed *	0	0	340	340	336
CW	1	0	322	323	319
Phone	3	0	213	216	215
Digital	16	0	126	142	142
160M	0	0	14	14	14
80M	3	0	129	132	131
40M	5	0	197	202	199
30M	13	0	162	175	174
20M	7	0	226	233	232
17M	13	0	175	188	187
15M	8	0	168	176	174
12M	11	0	131	142	141
10M	2	0	140	142	139
6M	0	0	5	5	5
Challenge *	62	0	1334	---	1396
5-Band *	---	---	---	---	---
5-Band 30M *	---	---	---	---	---
5-Band 17M *	---	---	---	---	---
5-Band 12M *	---	---	---	---	---

* = Award has been issued

I guess I need an antenna for 160 meters.

COMPUTER LOGGING PROGRAMS

Chasing DX, collecting confirmations, and applying for awards, requires a system to keep and manage data. Data management is what computers do best, and computer logging programs are the answer.

I am most familiar with DxLab Suite by AA6YQ, Dave, so that is the one I will describe here. Other logging programs may provide similar functionality. DxLab is free; there is a very active users group and a Wiki. Dave is tireless in providing advice and upgrades.

DxLab is not a contest logging program. Contest logging programs are specialized software written to keep tabs on contest exchanges and scoring. They do not operate to track QSLs and awards. Fortunately, contesting applications, such as N1MM+, will generate an ADIF file to import into DXLab. From DxLab, upload to OQRS, QDure, or LoTW and follow your overall progress.

DxLab is a suite of coordinated programs, so entering data in one populates the others. I will describe the function of each. There is no way I can be exhaustive in my explanation, and I will not try to tell you the "how." A suite of programs that does as much as DxLab has a learning curve. Don't despair. Visit the DxLab website and see what it has to offer. This explanation focuses on functionality.

Launcher is the gateway into the system. When Launcher starts, it checks for program and database updates and offers the option to download and install them. It then starts the various modules that make up the suite.

Commander connects the radio with the computer through a serial port, a USB to serial converter cable, or a USB port on both the radio and computer. The

COMPUTER LOGGING PROGRAMS

computer reads frequency and mode from the radio and automatically inserts that data into the log.

Commander can send instructions to the radio allowing direct frequency entry, mode, and filter changes. I prefer knobs to the mouse, so I minimize this window on my computer screen to save space.

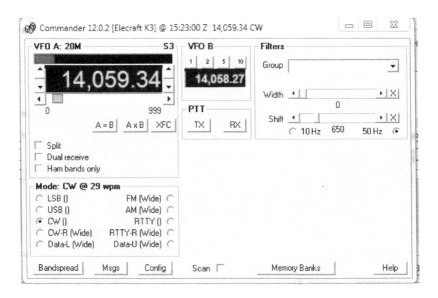

DxKeeper displays the logbook. DxKeeper can sort and filter, show the QSL status (sent, received, confirmed, or verified), print QSL labels, or upload your data to LoTW. The Check Progress tab will generate reports showing awards progress.

DxLab has a backup function that automatically backs up to DropBox every time you shut down DxLab, or on-demand preventing loss of valuable data.

Here is a screenshot from a particularly productive couple of days. That contact with Chesterfield Islands was a tough one. I had a hard time hearing them until an opening on 15 meters one evening. I was

COMPUTER LOGGING PROGRAMS

able to work both the SSB and CW stations within a few minutes of each other.

The shaded bands are yellow on my screen and indicate the station participates in LoTW.

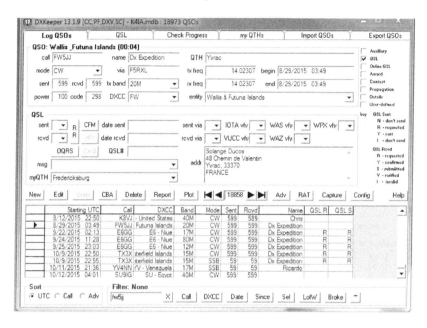

DxKeeper Capture is the QSO information entry box. Enter a call in the Capture box, and the system populates fields with information from the Internet or previous QSOs. Frequency and mode get imported through the Commander radio/computer interface. The date and time come from the computer. Hit the Log button to put the information in the logbook portion of DxKeeper, and the fields clear for your next QSO. You can't forget or accidentally enter the wrong information.

Notice the Capture window on the next page tells me the station participates in OQRS, LoTW, and eQSL. At the bottom, the Spot button enters the data in the SpotCollector module and broadcasts the information on spotting networks.

COMPUTER LOGGING PROGRAMS

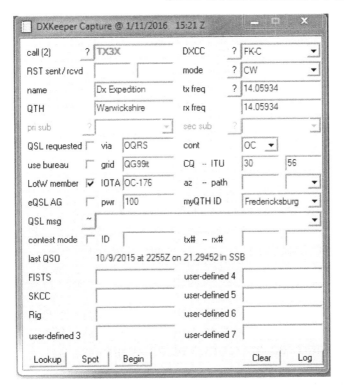

DxView is a chart showing the band and mode progress for that entity, distance, and bearing along with the time zone. There is also a World Map showing the short and long paths and grayline.

You can see the distance to the DX (8,887 miles short path), and clicking the ~ symbol gives you the long path distance. Clicking on the SP button will automatically turn your properly-equipped rotor to the short-path direction of the DX.

COMPUTER LOGGING PROGRAMS

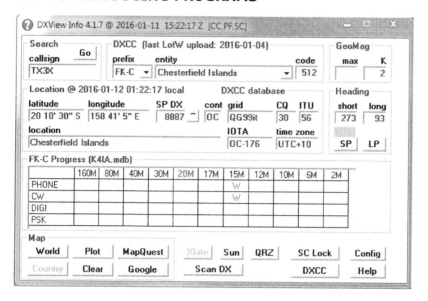

The red "W" in the chart tells me I have worked TX3 on 15-meter phone and CW, but I do not have the contacts confirmed or verified at this time.[31]

Pathfinder is the callbook lookup function. A callsign entered in the Capture box is exported to Pathfinder. Pathfinder looks up the station's data on QRZ.com, Buckmaster, and many other databases. It puts the operator's name and QTH into the DxKeeper Capture window (subject to your option to override) and displays the station's address and QSL route.

Pathfinder will lookup information the station has uploaded to QRZ.com, including, sometimes, the operator's picture. This screenshot shows the QRZ.com page for the Palmyra Islands DXpedition.

---

[31] A contact is "confirmed" when you get the QSL or LoTW confirmation. It is "verified" when submitted and accepted for DXCC credit.

COMPUTER LOGGING PROGRAMS

SpotCollector receives and aggregates data from multiple spotting sources. A Spotcollector screenshot is on the following page.

Three dots at the top of the display indicate a connection to three spotting sources. SpotCollecter alerts you to who is on, who uses LoTW, if they are working split, and who heard them. It can give audio and eMail alarms. "You've got DX!" I don't use that feature as it would drive me crazy.

Different color fonts indicate needed DX. Background colors identify LoTW and OQRS participants. Spotcollector will upload the spots you hear. Up-to-date solar flux and geomagnetic readings are in the upper left corner.

COMPUTER LOGGING PROGRAMS

SpotCollector will search for a callsign or entity to report the history of when a station was heard. Use that data to learn the operator's habits. If the DX is on the air around the same time every day, listen for him then. Spotcollector can display spots around your frequency and sort by band, mode, or entity.

Winwarbler is the digital interface supporting sending, receiving, and logging PSK, RTTY, and other Digital modes. There is a screenshot in the Modes Chapter.

PropView is a propagation prediction program. It collects current Solar and geomagnetic numbers from the Internet to develop a chart showing Maximum Usable Frequency at different confidence levels.

The following page displays a prediction for Palmyra & Jarvis Islands. Time of day is the bottom axis. The dark wavy line at the bottom is the Minimum Useable Frequency over time. The lighter lines above that are the Maximum Useable Frequencies at a 10%, 50%, and 90% confidence level. The dark vertical line is the present time. The short straight horizontal lines show predicted band openings, and it looks like 15 meters

COMPUTER LOGGING PROGRAMS

should be hopping right now. Unfortunately, I don't hear anything.

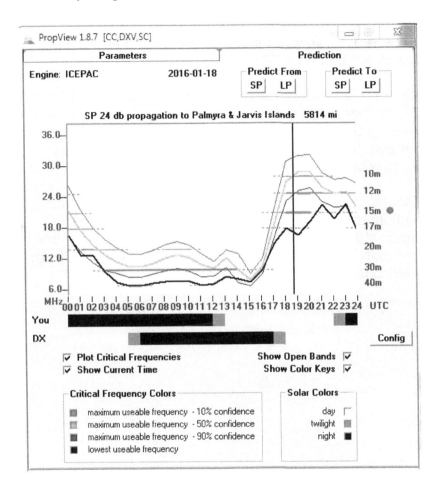

CONTESTS

"But, I am not a contester," you protest. No one said you had to be. Contests are a target-rich DX environment. Lots of stations are on the air, and they are all looking for QSOs.

Many DX operators contest and are anxious for all the QSOs they can muster. They will listen and try to dig you out because every QSO equals points for them. Serious contest stations answer QSLs because they want you to call them again in the next contest.

Contest clubs allow hams who are antenna or equipment challenged to band together and create super-stations members can operate. They are more common in Europe than in the US. Here's a card from a famous Russian contest station.

Contests provide an excellent opportunity to work numerous stations and entities, increasing your DX count quickly. Many contesters upload their logs to LoTW, so there are quick confirmations without sending a QSL card.

CONTESTS

How does a non-contester operate in a contest?

First, identify the contest. There are a dozen or so every weekend. Read the Contest Corral column in *QST* magazine. NG3K.com and EHam.net have lists of upcoming contests with links to the sponsor and rules. The smaller contests may offer suggested operating frequencies.

If you are looking for Asian stations, don't miss the All Asia contest sponsored by the Japanese Amateur Radio League in June. Need Africa? Look for the All Africa contest in September. There is a CQ World Wide contest in October (SSB) and November (CW).

ARRL has a CW DX contest in February and an SSB session in March. ARRL lists the objective as *"To encourage W/VE stations to expand knowledge of DX propagation on the HF and MF bands, improve operating skills, and improve station capability by creating a competition in which DX stations may only contact W/VE stations."* That means you are the fox!

Second, learn the exchange. What information does the contestant want? That will be on the contest website, or listen and figure it out. Typically it is a signal report (always 59) and serial number. Maybe it includes a CQ Zone Number, state, or power. None of that is difficult.

Third, get on the air. You are not trying to win, so don't get discouraged that someone sends a serial number in the thousands when you are just getting started. Your QSO #2 is just as valuable as QSO #1046. Plug away and watch your DXCC list grow.

In the end, send the log to the contest sponsor. Use an electronic upload from the computer logging program. Usually, it will convert the log to Cabrillo format first. That is a uniform contest format the

DX – The Easy Way

CONTESTS

sponsor uses to compare logs for errors. Your log helps verify the accuracy of the big guns who are competing. You are not trying to win, and no one will laugh at a low score. You never know, you may qualify for a certificate.

Here are a few contest operating tips:

Contests are about an exchange. Nothing more. Saying "Hi" to an old friend is OK, but no one wants to chit-chat or talk about the weather.

Contest exchanges are crisp and quick. Here are some examples of things not to say because they are unnecessary, repetitive, and slow down the pace:

Don't say "Roger the 59, number 1046."
He knows the exchange he gave. Ask for a repeat if you didn't copy, but there is no need to parrot it back.

Don't say, "Please copy."
Give the exchange; he already knows he has to copy.

Don't say, "When last heard you were 59."
What else would you be reporting? How he sounded before you last heard him? Just say "59."

Don't repeat the signal report. "599 599"
The exchange is always 599 on CW or 59, on phone. No need to repeat it. Use N for 9 on CW. "5NN."

Don't give his callsign again.
He already knows his callsign, and he knows you are calling him because he acknowledged you.

Don't give your callsign again.
He got it already, and if repeated, he will wonder if you are trying to make a correction.

AWARDS

If you like to collect wallpaper (certificates) and lumber (plaques), you picked the right hobby. There are hundreds of awards available.

The award I keep talking about is DXCC and its variants: the multiband DXCC and Honor Roll. Worthy endeavors, no doubt, but there are lots more.

ARRL also sponsors QRP (low power) DXCC, Worked All States, and Worked All Continents DX awards.

CQ magazine sponsors its version of DXCC. They also sponsor a Worked All Zones (WAZ) award for proof of contacts with all 40 CQ Zones. The WPX award recognizes the accomplishment of confirmed QSOs with the many prefixes throughout the world. All these awards have endorsements for mode and band.

Various national and international organizations offer awards for contacting member stations. Some require as few as a single contact. *CQ* magazine has a monthly column describing new awards from the world over, and there is also a wealth of information at DXAwards.com.

Special events are not awards but can net some fascinating QSL cards. I don't know much about Lithuanian history, but I found out they still commemorate the 600th anniversary of the Grunwald-Zalgiris battle fought during the Polish-Lithuanian-Teutonic War on July 15, 1410. The action was one of the largest in medieval Europe.

AWARDS

Special Event Station to commemorate the 600th anniversary of the Grunwald-Zalgiris battle.

I am too young to remember the formation of the State of Israel, but I was glad to help them celebrate their 50th anniversary. Mazeltov!

FISH STORIES – THE ONES THAT GOT AWAY

Sometimes, the fish slips off the hook. Sometimes, the fisherman isn't ready when the fish are biting. I offer the following fish stories to illustrate a few points.

BE READY

Be ready on as many bands and modes as possible. Here is the sad tale of an operator who was not. The opportunity arrived when 4L4FN, "Ed" from the Republic of Georgia, was stationed in North Korea with the United Nations World Food Program. In late 2001, after two years of begging, he got permission to operate an amateur radio station from the most enigmatic and elusive DXCC entity ever, the Hermit Kingdom.

Ed operated RTTY, and our sad operator, having just recently returned to the amateur ranks, was not set up to operate RTTY. He kept thinking Ed would undoubtedly tire of listening to the diddles of RTTY and get on CW. It never happened.

In November 2002, Ed was abruptly ordered off the air by the North Korean government. Ed made more than 16,000 contacts, but the humbled operator was not among them. I guess you figured out who that humbled operator was, de K4IA. I've been waiting 20 years for another P5 activation. Rumors abound, but the prospects are dim. This time, I'll be ready.

FISH STORIES

BE QUICK

You've heard a DXpedition is going to the end of the world and will operate for two weeks. When first heard, the pileups are thick, so you decide to wait a few days and let the crowds thin out. Surely, it will be easier later, and it may be. Or, maybe the rest of the world hasn't woken up to the activation, and the crowds will grow. You never know, and you also don't know what might intervene to interrupt the DXpedition.

In 2004, a team went to Andaman and Nicobar Islands (VU4), located in the Bay of Bengal off the coast of Myanmar. VU4 is in the Very Rare DX category. Reports at the time rated it as rare as North Korea. Very rare, not only because it required special permission from the Indian government, but the path from North America was almost 9,000 miles over the North Pole. Over-the-pole signals are subject to auroral disruption and absorption. The operation was "mighty lite," meaning not a lot of equipment, amplifiers, antennas, seasoned operators, etc. They were tough to hear in North America.

On December 26, an earthquake and resulting tsunami claimed several hundred lives and devastated the islands. The DXpedition went into disaster mode and provided the only emergency communications available to the mainland. Needless to say, they stopped taking calls from the pileups.

Earthquakes and tsunamis are rare, but all kinds of things can go wrong to derail a DXpedition. These include medical emergencies, solar storms, equipment failures, and extreme weather. The moral of the story is, "Don't wait." Get them at least once, early, and come back to fill in other bands and modes later if you must. Recognize that it may be a decade or longer before a Very Rare DX is on the air again.

FISH STORIES

The VU4 team performed their disaster duties so well the Indian government reversed its restrictive policy on amateur activity from VU4. The "mighty lite" crew showed Amateur Radio works when all else fails. There have been a few more DXpeditions since. I finally nailed one in 2014, but it wasn't easy.

Here is a press release from the recent DXpedition to the South Sandwich Islands with another example:
NEWS UPDATE @ 21:20Z - Nigel Jolly, Braveheart owner, declared an emergency and ordered the VP8STI team to cease all operations and return to the ship. A large ice flow that had broken away due to last night's storm starts to block the entrance to the Bay where they were camped. There was strong potential for the ice flow to prevent Braveheart reaching them.

The Team is now safely aboard the Braveheart and they are moving away from Southern Thule Is. All equipment and personal gear is still on the island. They hope to return to camp site during next good weather window. Once they retrieve their equipment they will make a determination as to whether they can proceed to South Georgia to do VP8SGI as planned - VP8STI QRT !

FISH STORIES

BE CURRENT

You will become familiar with the prefixes used to identify various countries, but unusual prefixes are not easily recognized. We all know that "F" identifies a station in France. But, "TO" can also designate a French station.

Logging programs use database files to keep up with odd prefixes. Be sure to keep the database files current. This sometimes requires a manual override, such as when K5K operated from Kingman Reef or K7C from Kure Island. Both look like US special-event stations, and your logging program would probably report them as such.

In 2003, TO4E operated from Very Rare Juan de Nova/Europa Island near Madagascar. Europa requires special permission to visit and is very difficult to access because of steep cliffs and rough seas, so it is rarely activated. I heard them several times, but my logging program reported the "TO" prefix as France. I assumed it was some French special-event station. Who wants to fight a pileup for that? I wasn't interested.

I did not have my database up to date, and I had not paid attention to the announcements — big mistakes. Europa is one I could have caught, but I let it slip off the hook. It took 13 years before anyone went back. I got them the second time.

The moral of this story is to pay attention to the DX bulletins and keep your databases current. If you hear a massive pileup, work it first and ask questions later. That's Rule #1 – work first, verify later.

FISH STORIES

Malpelo is 310 miles west of Colombia and uninhabited except for a small military post. Imagine scaling those cliffs to set up your station. Is this how Europa looks?

Tristan Da Cunha seems a bit more inviting. Named after its Portuguese discoverer, TDC lies 1,500 miles west of South Africa and is the world's most remote inhabited (pop 267) archipelago. The main settlement is Edinburgh of the Seven Seas.

LEARNING CW

CW should not be that difficult. Think of it as a foreign language with 26 words and count to 10. Would you think it an impossible task? Probably not. I learned Morse code, and I can't memorize anything. Thankfully, you aren't memorizing. You are training to recognize a sound.

Boy Scouts was my first exposure to CW. I had to relearn it as a Ham. What went wrong?

In Boy Scouts, we memorized CW as dots and dashes. The decoding was visual, not audible. You had a little chart and looked up every letter. What's wrong with that? Your brain has to translate what it sees or hears into dots and dashes and then translate those into letters. That is a multi-step process. It is like learning to translate from English to Spanish to get to French.

I thought I understood it pretty well until someone sent Morse code by flashlight – blinking light to dots and dashes to letters. It was even worse when sent by waving a flag. I was lost.

Radio CW is audible. Recognize the sound, not memorize dots and dashes. "A" is not short-long or dot-dash. It is not even dit-dah. It is the sound made by dit-dah. Learning the sound eliminates all the in-between translations.

For the same reason, do not learn that "A" sounds like "Ah-pull" or that the letter "A" has a short line and a long line. I've seen pictograms like a child's alphabet block with a picture of a bee for "B." These gimmicks introduce additional mental steps. Now you are going from English to Spanish to French to get to Russian.

Learn CW by hearing it one or two letters at a time until you can immediately make the connection from

LEARNING CW

the sound to the letter. Then go on to the next letters and build. This is called the Koch method.

Another impediment to my learning was the way we sent CW. At slow speeds, "A" became the sound made by diiiiiiit-daaaaaaaaaah. Then, the next letter came immediately with no time for the brain to work. The modern way is called Farnsworth timing.

Farnsworth timing is to send the letters at 16-20 words per minute,[32] so each letter has one distinct sound. Individual dits and dahs do not form a letter. Don't try to hear the different elements. The letter is one sound.

To slow down the pace, Farnsworth increases the space between letters and words. Increasing the space does two things. It reinforces a single sound as the letter, and it gives the brain extra time between letters to work the translation. Also, learn to send using Farnsworth timing. That is how the other guy learned to receive. I set my keyer around 22-26 words per minute and slow down by increasing spaces.

After recognizing the letters, the next step is to recognize words. When reading, you do not see letters; but words. "Word" is not "W-O-R-D." The same happens with Morse code. Learn to recognize your callsign, RST, 5NN, TU, 73, and other common "words" without thinking about the individual letters or the elements that make up the letters.

Learning Code takes practice. In the beginning, listen to a Code practice CD or audio file. K7QO offers a free course download on his website, K7QO.net. G4FON has a Koch trainer at G4FON.net. There are other sources, as well.

---

[32] Words per minute (WPM) is based on a 5 letter word. "Paris" is an often-used standard. Send "Paris" 20 times in a minute for 20 WPM.

LEARNING CW

Understand the letters (mostly), then listen to some QSOs to learn the standard QSO protocol. Then, GET ON THE AIR! There is no better practice than making actual contacts. Real QSOs are exciting and won't seem like tedious practice.

FISTS CW Club and Straight Key Century Club (SKCC) promote CW, and there is slow CW on the frequencies suggested on their web pages. Both FISTS CW Club and SKCC assign a member number to exchange with other members to collect awards. It is a fun challenge.

Suggested frequencies to find slower CW include:
3.550 – 3.570 MHz
7.055 – 7.060 MHz
14.055 – 14.060 MHz
21.055 – 21.060 MHz
28.055 – 28.060 MHz
The old Novice Bands also have slow CW.

At first, concentrate on the QSO Trinity. Learn the pattern and know what to expect. With time, venture into more complex conversations. Try to match the speed of the other guy, but if you can't, "QRS" means "slow down" and "QRQ" means "speed up."

For more serious practice, tune to the W1AW Code Practice Sessions. These are texts from *QST* magazine, and because the words are longer and not always as predictable, harder to copy. ARRL offers code proficiency certificates that will look beautiful on your wall.

For the ultimate challenge, jump into a CW contest. Contests are almost guaranteed to add 5-10 WPM to your speed.

Here's my HF/VHF/UHF mobile rig installed in a Jaguar XJ6. The radio is in the trunk. A control head mounts

LEARNING CW

in what used to be the ashtray. The Morse code paddles are to the left of the gear shift. Do you think I like CW?

This is what I call "the cat's meow."

SUMMARY & A FINAL THOUGHT

This book promised to show The Easy Way to chase, work, and confirm DX contacts. Here is a review of The Easy Way methods:

- Be available on all bands and modes
- Listen, listen and listen some more
- Call CQ
- Learn enough CW to recognize your call and be able to decipher the DX call one letter at a time
- Use a voice and CW keyer to call CQ or send your callsign and 5NN on CW
- Use spotting networks
- Use a panadapter
- Upgrade your antenna first, then upgrade your power with a mid-range amplifier or 2-element beam
- Work contests
- Use LoTW, OQRS, and GlobalQSL
- Use a computer logging program
- Most importantly, follow the DX Code of Conduct that it may go well with you, and your DXCC total may multiply greatly.

Final thought: Anyone who listens to DX will see that pileup behavior has deteriorated. The recent chaos and DQRM (deliberate interference) around the Palmyra and South Sandwich DXpeditions was among the worst I have heard. Technology has brought many improvements to equipment. Now should be the Golden Age. Yet, our DX enjoyment is being destroyed by bad manners, greed, and stupidity. I hope that this book will open eyes and restore civility. Please, do your part.

73/DX
Buck
K4ia

DX ENTITY LIST

Following is a list of entities and prefixes sorted by prefix. Some of these prefixes are rarely used. The "Root" is the primary prefix you will see most often.

Prefix	Root	Entity
1A	1A	Sovereign Military Order of Malta
2A-2Z	G	England
2D	GD	Isle of Man
2I	GI	No Ireland
2J	GJ	Jersey
2M	GM	Scotland
2U	GU	Guernsey
2W	GW	Wales
3A	3A	Monaco
3B	3B8	Mauritius
3B7	3B6	Agalega & St Brandon
3B9	3B9	Rodriguez Is
3C	3C	Equatorial Guinea
3C0	3C0	Annobon Is
3D2	3D2	Conway Reef
3D2	3D2	Rotuma
3D6	3DA	Swaziland
3DA-M	3DA	Swaziland
3DN	3D2	Fiji
3DZ	3D2	Fiji
3E-3F	HP	Panama
3G	CE	Chile
3H-3U	BY	China
3V	3V	Tunisia
3W	3W	Vietnam
3X	3X	Guinea
3Y	3Y	Bouvet
3Y	3Y	Peter I
3Y	CE9	Antarctica

DX ENTITY LIST

Prefix	Root	Entity
3Y	LA	Norway
3Z	SP	Poland
4A-4C	XE	Mexico
4D-4I	DU	Philippines
4J1	R1MV	Malyj Vysotskij Is
4J-4K	4J	Azerbaijan
4K1	CE9	Antarctica
4K1	VP8	So Shetland Is
4K2	R1FJ	Franz Josef Land
4L	4L	Georgia
4M	YV	Venezuela
4N5	Z3	Macedonia
4O	4O	Montenegro
4P-4S	4S	Sri Lanka
4T	OA	Peru
4U	4U	ITU Geneva
4U	4U	UN HQ
4V	HH	Haiti
4W	4W	Timor-Leste
4X	4X	Israel
4Y	4Y	(ICAO)
4Z	4X	Israel
5A	5A	Libya
5B	5B	Cyprus
5C-5G	CN	Morocco
5H-5I	5H	Tanzania
5J-5K	HK	Colombia
5L-5M	EL	Liberia
5N-5O	5N	Nigeria
5P-5Q	OZ	Denmark
5R-5S	5R	Madagascar
5T	5T	Mauritania
5U	5U	Niger
5V	5V	Togo
5W	5W	Samoa
5X	5X	Uganda

DX ENTITY LIST

Prefix	Root	Entity
5Y-5Z	5Z	Kenya
6A-6B	SU	Egypt
6C	YK	Syria
6D-6J	XE	Mexico
6K-6N	HL	South Korea
6O	T5	Somalia
6P-6S	AP	Pakistan
6T-6U	ST	Sudan
6V-6W	6W	Senegal
6X	5R	Madagascar
6Y	6Y	Jamaica
6Z	EL	Liberia
7A-7I	YB	Indonesia
7J-7N	JA	Japan
7O	7O	Yemen
7P	7P	Lesotho
7Q	7Q	Malawi
7R	7X	Algeria
7S	SM	Sweden
7T-7Y	7X	Algeria
7X	7X	Algeria
7Z	HZ	Saudi Arabia
8A-8I	YB	Indonesia
8J1	CE9	Antarctica
8J-8N	JA	Japan
8O	A2	Botswana
8P	8P	Barbados
8Q	8Q	Maldives
8R	8R	Guyana
8S	SM	Sweden
8T-8Y	VU	India
8Z	HZ	Saudi Arabia
9A	9A	Croatia
9B-9D	EP	Iran
9E-9F	ET	Ethiopia
9G	9G	Ghana

DX ENTITY LIST

Prefix	Root	Entity
9H	9H	Malta
9I-9J	9J	Zambia
9K	9K	Kuwait
9L	9L	Sierra Leone
9M	9M2	W Malaysia
9M2	9M2	W Malaysia
9M4	9M2	W Malaysia
9M6	9M6	E Malaysia
9M8	9M6	E Malaysia
9MO	1S	Spratly Is
9N	9N	Nepal
9O-9T	9Q	Dem Rep Congo (Zaire)
9U	9U	Burundi
9V	9V	Singapore
9W	9M2	W Malaysia
9X	9X	Rwanda
9Y-9Z	9Y	Trinidad & Tobago
A2	A2	Botswana
A3	A3	Tonga
A4	A4	Oman
A5	A5	Bhutan
A6	A6	United Arab Emirates
A7	A7	Qatar
A8	EL	Liberia
A9	A9	Bahrain
AA-AL	K	United States
AH0	KH0	Mariana Is
AH1	KH1	Baker & Howland Is
AH2	KH2	Guam
AH3	KH3	Johnston Is
AH4	KH4	Midway Is
AH5	KH5	Palmyra, Jarvis Is
AH5K	KH5K	Kingman Reef
AH6 AH7	KH6	Hawaii
AH7K	KH7K	Kure Is
AH8	KH8	American Samoa

DX ENTITY LIST

Prefix	Root	Entity
AH8S	KH8S	Swain's Island
AH9	KH9	Wake Is
AL	KL	Alaska
AM-AO	EA	Spain
AP-AS	AP	Pakistan
AT0	CE9	Antarctica
AT-AW	VU	India
AX	VK	Australia
AY-AZ	LU	Argentina
BA-BZ	BY	China
BM-BQ	BV	Taiwan
BS7	BS7	Scarborough Reef
BU	BV9	Pratas Is
BV	BV	Taiwan
BV9	BV9	Pratas Is
C2	C2	Nauru
C3	C3	Andorra
C4	5B	Cyprus
C5	C5	Gambia
C6	C6	Bahamas
C7	C7	(WMO)
C8-C9	C9	Mozambique
CA-CE	CE	Chile
CE0X	CE0X	San Felix
CE0Y	CE0Y	Easter Is
CE0Z	CE0Z	Juan Fernandez
CE9	CE9	Antarctica
CE9	VP8	So Shetland Is
CF-CK	VE	Canada
CL-CM	CO	Cuba
CN	CN	Morocco
CO	CO	Cuba
CP	CP	Bolivia
CQ-CU	CT	Portugal
CT3	CT3	Madeira Is
CU	CU	Azores

DX – The Easy Way

DX ENTITY LIST

Prefix	Root	Entity
CV-CX	CX	Uruguay
CY0	CY0	Sable Is
CY9	CY9	St Paul Is
CY-CZ	VE	Canada
D2-D3	D2	Angola
D4	D4	Cape Verde
D5	EL	Liberia
D6	D6	Comoros
D7-D9	HL	So Korea
DA-DR	DL	Germany
DPO	CE9	Antarctica
DS-DT	HL	So Korea
DU	DU	Philippines
DZ	DU	Philippines
E2	HS	Thailand
E3	E3	Eritrea
E4	E4	Palestine
E5	E5	So Cook Is
E5	E5	No Cook Is
E6	E6	Niue
E7	E7	Bosnia-Hercegovina
EA6	EA6	Balearic Is
EA8	EA8	Canary Is
EA9	EA9	Ceuta & Melilla
EA-EH	EA	Spain
EI-EJ	EI	Ireland
EK	EK	Armenia
EL	EL	Liberia
EM-EO	UR	Ukraine
EP-EQ	EP	Iran
ER	ER	Moldova
ES	ES	Estonia
ET	ET	Ethiopia
ET2	E3	Eritrea
EU-EW	EV	Belarus
EX	EX	Kyrgyzstan

DX ENTITY LIST

Prefix	Root	Entity
EY	EY	Tajikistan
EZ	EZ	Turkmenistan
F	F	France
FA-FZ	F	France
FG	FG	Guadeloupe
FH	FH	Mayotte
FJ	FJ	St Barthelemy
FK	FK	Chesterfield Is
FK	FK	New Caledonia
FM	FM	Martinique
FO	FO	French Polynesia
FO0	FO0	Austral Is
FO0	FO0	Clipperton Is
FO0	FO0	Marquesas Is
FP	FP	St Pierre & Miquelon
FR	FR	Reunion
FR/G	FT_G	Glorioso Is
FR/J	FT_J	Juan de Nova, Europa
FR/T	FT_T	Tromelin
FS	FS	St Martin
FT_E	FT_J	Juan de Nova, Europa
FT_J	FT_G	Glorioso Is
FT_J	FT_J	Juan de Nova, Europa
FT_T	FT_T	Tromelin
FT_W	FT_W	Crozet
FT_X	FT_X	Kerguelen Is
FT_Y	CE9	Antarctica
FT_Z	FT_Z	Amsterdam & St Paul Is
FW	FW	Wallis & Futuna Is
FY	FY	French Guiana
G	G	England
GA-GZ	G	England
GC	GW	Wales
GD	GD	Isle of Man
GH	GJ	Jersey
GI	GI	No Ireland

DX ENTITY LIST

Prefix	Root	Entity
GJ	GJ	Jersey
GM	GM	Scotland
GN	GI	No Ireland
GP	GU	Guernsey
GS	GM	Scotland
GT	GD	Isle of Man
GU	GU	Guernsey
GW	GW	Wales
GZ	GM	Scotland
H2	5B	Cyprus
H3	HP	Panama
H4	H4	Solomon Is
H40	H40	Temotu
H5	ZS	South Africa
H6-H7	YN	Nicaragua
H8-H0	HP	Panama
HA	HA	Hungary
HB	HB	Switzerland
HB0	HB0	Liechtenstein
HC8	HC8	Galapagos Is
HC-HD	HC	Ecuador
HE	HB	Switzerland
HF	SP	Poland
HF0	VP8	So Shetland Is
HG	HA	Hungary
HH	HH	Haiti
HI	HI	Dominican Rep
HJ-HK	HK	Colombia
HK0	HK0	Malpelo Is
HK0	HK0	San Andres & Providencia
HL	HL	South Korea
HM	P5	North Korea
HN	YI	Iraq
HO-HP	HP	Panama
HP	HP	Panama
HQ-HR	HR	Honduras

DX ENTITY LIST

Prefix	Root	Entity
HS	HS	Thailand
HT	YN	Nicaragua
HU	YS	El Salvador
HU	YS	El Salvador
HV	HV	Vatican
HW-HY	F	France
HZ	HZ	Saudi Arabia
I	I	Italy
IA-IZ	I	Italy
IS	IS	Sardinia
J2	J2	Djibouti
J3	J3	Grenada
J4	SV	Greece
J5	J5	Guinea-Bissau
J6	J6	St Lucia
J7	J7	Dominica
J8	J8	St Vincent
JA	JA	Japan
JA-JS	JA	Japan
JD	JD	Minami Torishima
JD	JD	Ogasawara
JT-JV	JT	Mongolia
JW	JW	Svalbard
JW-JX	LA	Norway
JX	JX	Jan Mayen
JY	JY	Jordan
JZ	YB	Indonesia
K	K	United States
KA-KZ	K	United States
KC4	CE9	Antarctica
KC6	T8	Palau
KC6	V6	Micronesia
KG4	KG4	Guantanamo Bay
KH0	KH0	Mariana Is
KH1	KH1	Baker & Howland Is
KH2	KH2	Guam

DX ENTITY LIST

Prefix	Root	Entity
KH3	KH3	Johnston Is
KH4	KH4	Midway Is
KH5	KH5	Palmyra, Jarvis Is
KH5K	KH5K	Kingman Reef
KH6 KH7	KH6	Hawaii
KH7K	KH7K	Kure Is
KH8	KH8	Am Samoa
KH8S	KH8S	Swain's Island
KH9	KH9	Wake Is
KL	KL	Alaska
KP1	KP1	Navassa Is
KP2	KP2	Virgin Is
KP3	KP4	Puerto Rico
KP4	KP4	Puerto Rico
KP5	KP5	Desecheo Is
KX6	V7	Marshall Is
L2-L9	LU	Argentina
LA	LA	Norway
LA-LN	LA	Norway
LO-LW	LU	Argentina
LU	LU	Argentina
LU_Z	CE9	Antarctica
LU_Z	VP8	So Georgia Is
LU_Z	VP8	So Orkney Is
LU_Z	VP8	So Sandwich Is
LU_Z	VP8	So Shetland Is
LX	LX	Luxembourg
LY	LY	Lithuania
LZ	LZ	Bulgaria
MA-MZ	G	England
MC	GW	Wales
MD	GD	Isle of Man
MH	GJ	Jersey
MI	GI	No Ireland
MJ	GJ	Jersey
MM	GM	Scotland

DX ENTITY LIST

Prefix	Root	Entity
MN	GI	No Ireland
MP	GU	Guernsey
MS	GM	Scotland
MT	GD	Isle of Man
MU	GU	Guernsey
MW	GW	Wales
MZ	GM	Scotland
NA-NZ	K	United States
NH0	KH0	Mariana Is
NH1	KH1	Baker & Howland Is
NH2	KH2	Guam
NH3	KH3	Johnston Is
NH4	KH4	Midway Is
NH5	KH5	Palmyra, Jarvis Is
NH5K	KH5K	Kingman Reef
NH6 NH7	KH6	Hawaii
NH7K	KH7K	Kure Is
NH8	KH8	Am Samoa
NH8S	KH8S	Swain's Island
NH9	KH9	Wake Is
NL	KL	Alaska
NP3	KP4	Puerto Rico
NP4	KP4	Puerto Rico
NP5	KP5	Desecheo Is
OA-OC	OA	Peru
OD	OD	Lebanon
OE	OE	Austria
OF-OJ	OH	Finland
OH	OH	Finland
OH0	OH0	Aland Is
OJ0	OJ0	Market Reef
OK-OL	OK	Czech Republic
OM	OM	Slovakia
ON-OT	ON	Belgium
OU-OZ	OZ	Denmark
OW	OY	Faroe Is

DX ENTITY LIST

Prefix	Root	Entity
OX	OX	Greenland
OY	OY	Faroe Is
OZ	OZ	Denmark
P2	P2	Papua New Guinea
P3	5B	Cyprus
P4	P4	Aruba
P5	P5	North Korea
P5-P9	P5	North Korea
PA-PJ	PA	Netherlands
PJ2	PJ2	Curacao
PJ4	PJ4	Bonaire
PJ5	PJ5	Saba & St Eustatius
PJ6	PJ5	Saba & St Eustatius
PJ7	PJ7	St Maarten
PK-PO	YB	Indonesia
PP-PY	PY	Brazil
PY0F	PY0F	Fernando de Noronha
PY0P	PY0P	St Peter & St Paul Rocks
PY0T	PY0T	Trindade & Martin Vaz Is
PY0ZF	PY0F	Fernando de Noronha
PY0ZP	PY0P	St Peter & St Paul Rocks
PY0ZT	PY0T	Trindade & Martin Vaz Is
PZ	PZ	Surinam
R1AN	CE9	Antarctica
R1AN	VP8	So Shetland Is
R1FJ	R1FJ	Franz Josef Land
R1MV	R1MV	Malyj Vysotskij Is
RA-RZ	UA	Russia
S0~	S0	Western Sahara
S2-S3	S2	Bangladesh
S4	ZS	South Africa
S5	S5	Slovenia
S6	9V	Singapore
S7	S7	Seychelles
S8	ZS	South Africa
S9	S9	Sao Tome & Principe

DX ENTITY LIST

Prefix	Root	Entity
SA-SM	SM	Sweden
SM	SM	Sweden
SN-SR	SP	Poland
SP	SP	Poland
SSA-SSM	SU	Egypt
SSN-SSZ	ST	Sudan
ST	ST	Sudan
SU	SU	Egypt
SV1/A	SV1/A	Mt Athos
SV5	SV5	Dodecanese
SV9	SV9	Crete
SV-SZ	SV	Greece
SY2	SV1/A	Mt Athos
T2	T2	Tuvalu
T30	T30	West Kiribati
T31	T31	Central Kiribati
T32	T32	East Kiribati
T33	T33	Banaba
T4	CO	Cuba
T5	T5	Somalia
T6	T6	Afghanistan
T7	T7	San Marino
T8	T8	Palau
T9	E7	Bosnia-Hercegovina
TA-TC	TA	Turkey
TD	TG	Guatemala
TE	TI	Costa Rica
TF	TF	Iceland
TG	TG	Guatemala
TH	F	France
TI	TI	Costa Rica
TI9	TI9	Cocos Is
TJ	TJ	Cameroon
TK	F	France
TK	TK	Corsica
TL	TL	Central African Rep

DX ENTITY LIST

Prefix	Root	Entity
TM	F	France
TN	TN	Congo
TO	FM	Martinique
TO	FR	Reunion
TO	FS	St Martin
TO	FT_G	Glorioso Is
TO	FT_J	Juan de Nova, Europa
TO	FT_T	Tromelin
TO	FT_W	Crozet
TO	FY	French Guiana
TO-TQ	F	France
TO-TQ	FG	Guadeloupe
TR	TR	Gabon
TS	3V	Tunisia
TT	TT	Chad
TU	TU	Ivory Coast
TV-TX	F	France
TX	FH	Mayotte
TX	FK	Chesterfield Is
TX	FK	New Caledonia
TX	FO	French Polynesia
TX	FO0	Austral Is
TX	FO0	Clipperton Is
TX	FO0	Marquesas Is
TX	FP	St Pierre & Miquelon
TX	FT_X	Kerguelen Is
TX	FT_Z	Amsterdam & St Paul Is
TX	FW	Wallis & Futuna Is
TY	TY	Benin
TZ	TZ	Mali
U	UA	Russia
UA	UA	Russia
UA2	UA2	Kaliningrad
UA9, UA0	UA9	Russia (Asiatic)

DX ENTITY LIST

Prefix	Root	Entity
UA-UI	UA	Russia
UB	UR	Ukraine
UC	EV	Belarus
UD	4J	Azerbaijan
UF	4L	Georgia
UG	EK	Armenia
UI	UK	Uzbekistan
UJ	EY	Tajikistan
UJ	EZ	Turkmenistan
UJ-UM	UK	Uzbekistan
UK	UK	Uzbekistan
UL	UN	Kazakhstan
UM	EX	Kyrgyzstan
UN-UQ	UN	Kazakhstan
UO	ER	Moldova
UP	LY	Lithuania
UP	OA	Peru
UQ	YL	Latvia
UR	ES	Estonia
UR-UZ	UR	Ukraine
V2	V2	Antigua, Barbuda
V3	V3	Belize
V4	V4	St Kitts, Nevis
V5	V5	Namibia
V6	V6	Micronesia
V7	V7	Marshall Is
V8	V8	Brunei
V9	ZS	South Africa
VA-VG	VE	Canada
VE	VE	Canada
VH-VN	VK	Australia
VK	VK	Australia
VK0	CE9	Antarctica
VK0	VK0	Heard Is
VK0	VK0	Macquarie Is
VK9C	VK9C	Cocos-Keeling Is

DX ENTITY LIST

Prefix	Root	Entity
VK9C	VK9W	Willis Is
VK9L	VK9L	Lord Howe Is
VK9M	VK9M	Mellish Reef
VK9N	VK9N	Norfolk Is
VK9W	VK9W	Willis Is
VK9X	VK9X	Christmas Is
VK9Y	VK9C	Cocos-Keeling Is
VK9Z	VK9M	Mellish Reef
VP2E	VP2E	Anguilla
VP2M	VP2M	Montserrat
VP2V	VP2V	Br Virgin Is
VP5	VP5	Turks & Caicos Is
VP6	VP6	Pitcairn Is
VP6	VP6	Ducie Is
VP8	CE9	Antarctica
VP8	VP8	Falkland Is
VP8	VP8	So Georgia Is
VP8	VP8	So Orkney Is
VP8	VP8	So Sandwich Is
VP8	VP8	So Shetland Is
VP9	VP9	Bermuda
VP-VQ	G	England
VQ9	VQ9	Chagos
VR	BY	China
VR2	VR2	Hong Kong
VR6	VP6	Pitcairn Is
VS	G	England
VS6	VR2	Hong Kong
VT-VW	VU	India
VU	VU	India
VU4	VU4	Andaman & Nicobar Is
VU7	VU7	Lakshadweep Is
VX-VY	VE	Canada
VZ	VK	Australia
WA-WZ	K	United States
WH0	KH0	Mariana Is

DX ENTITY LIST

Prefix	Root	Entity
WH1	KH1	Baker & Howland Is
WH2	KH2	Guam
WH3	KH3	Johnston Is
WH4	KH4	Midway Is
WH5	KH5	Palmyra, Jarvis Is
WH5K	KH5K	Kingman Reef
WH6-7	KH6	Hawaii
WH7K	KH7K	Kure Is
WH8	KH8	Am Samoa
WH8S	KH8S	Swain's Island
WH9	KH9	Wake Is
WL	KL	Alaska
WP1	KP1	Navassa Is
WP2	KP2	Virgin Is
WP3	KP4	Puerto Rico
WP4	KP4	Puerto Rico
WP5	KP5	Desecheo Is
XA-XI	XE	Mexico
XE	XE	Mexico
XF4	XF4	Revilla Gigedo
XJ-XO	VE	Canada
XP	OX	Greenland
XP	OZ	Denmark
XQ-XR	CE	Chile
XS	BY	China
XT	XT	Burkina Faso
XU	XU	Cambodia
XV	3W	Vietnam
XV9	1S	Spratly Is
XW	XW	Laos
XX	CT	Portugal
XX9	XX9	Macao
XY-XZ	XZ	Myanmar (Burma)
XZ	XZ	Myanmar (Burma)
Y2	DL	Germany
Y9	DL	Germany

DX – The Easy Way

DX ENTITY LIST

Prefix	Root	Entity
YA	T6	Afghanistan
YB	YB	Indonesia
YB-YH	YB	Indonesia
YI	YI	Iraq
YJ	YJ	Vanuatu
YK	YK	Syria
YL	YL	Latvia
YM	TA	Turkey
YN	YN	Nicaragua
YO-YR	YO	Romania
YS	YS	El Salvador
YT-YU	YU	Serbia
YU3	4O	Montenegro
YU6	4O	Montenegro
YV0	YV0	Aves Is
YV-YY	YV	Venezuela
Z2	Z2	Zimbabwe
Z3	Z3	Macedonia
Z8	Z8	South Sudan
ZA	ZA	Albania
ZB	ZB	Gibraltar
ZB-ZJ	G	England
ZC	ZC	Cyprus SBA
ZD7	ZD7	St Helena
ZD8	ZD8	Ascension Is
ZD9	ZD9	Tristan da Cunha & Gough Is
ZF	ZF	Cayman Is
ZK1	E5	North Cook Is
ZK1	E5	South Cook Is
ZK3	ZK3	Tokelau Is
ZK7	ZL7	Chatham Is
ZK-ZM	ZL	New Zealand
ZL5	CE9	Antarctica
ZL8	ZL8	Kermadec Is
ZL9	ZL9	NZ Subarctic Is
ZN-ZO	G	England

DX ENTITY LIST

Prefix	Root	Entity
ZP	ZP	Paraguay
ZQ	GD	Isle of Man
ZR-ZU	ZS	South Africa
ZS	ZS	South Africa
ZS7	CE9	Antarctica
ZS8	ZS8	Pr Edward & Marion Is
ZV-ZZ	PY	Brazil
ZX0	CE9	Antarctica

Made in the USA
Middletown, DE
18 February 2022